Inorganic Chemistry Concepts
Volume 12

F. Valach · J. Ondráček · M. Melník

Crystallographic Statistics in Chemical Physics

An Approach to Statistical Evaluation
of Internuclear Distances in Transition
Element Compounds

With 82 Figures

Springer-Verlag Berlin Heidelberg GmbH

Dr. Fedor Valach
Department of Chemical Physics and Nuclear Technique
Dr. Milan Melník
Department of Inorganic Chemistry
Slovak Technical University
ČSSR — 81237 Bratislava

Dr. Jan Ondráček
Institute of Chemical Technology
Computing Centre
ČSSR — 16628 Prague-6

ISBN 978-3-662-01601-5   ISBN 978-3-662-01599-5 (eBook)
DOI 10.1007/978-3-662-01599-5

Library of Congress Cataloging-in-Publication Data
Valach, F. (Fedor), 1946–
Crystallographic statistics in chemical physics : an approach to statistical evaluation of inter-
nuclear distances in transition element compounds / F. Valach, J. Ondráček, M. Melník.
p.   cm. — (Inorganic chemistry concepts ; v. 12) Includes bibliographies and index.
ISBN 978-3-662-01601-5
1. Crystallography, Mathematical. 2. Transition metal compounds. 3. Crystallography—
Statistical methods. I. Ondráček, J. (Jan). 1955– . II. Melník, M. (Milan), 1938– . III. Title.
IV. Series.
QD921.V32 1988 88-4451
546'.6—dc19

© Springer-Verlag Berlin Heidelberg 1988
Originally published by Springer-Verlag Berlin Heidelberg New York in 1988
Softcover reprint of the hardcover 1st edition 1988

Offsetprinting: Color-Druck, Berlin;

2152/3020-543210

"The tendency towards symmetry is one
of the greatest law of inorganic nature"

E. Mallard

"It is the dissymetry which creates
the phenomenon"

P. Curie

Preface

The present development of crystal structure information systems cumulates data of some ten-thousand crystal structures of compounds. The annual addition of data (several thousand) indicates a vehement increase of the volume of crystal structural data, even of such groups of compounds which some ten years ago were studied only rarely. These data bases give the possibility to develop and apply techniques to search for the most general laws and rules of structures in condensed matter using statistical approach.

For the most part of "structure" chemists statistical approach in the region of solving problems of structures and bondings in crystals is not very familiar. The use of data bases of structures, however, makes its application in decision making processes unavoidable. Some statistical methods were already applied in the work by Pauling "Nature of the Chemical Bonds" to which mostly chemists referred, and they led to several useful concepts suitable for interpretation of results of crystal structural research of some more simple, for the most part of inorganic and organometallic compounds. For many inorganic and organometallic compounds, however, these concepts usually are not applicable. Nevertheless we may expect that statistical analysis of great sets of crystal structure data will allow to determine the limits of applicability of already accustomed generalizations, as well as creating new concepts suitable for interpreting and foreseeing crystal structures also of chemically more complicated systems.

Application of statistical methods is suitable when there are noisy data impeded by effects of other factors not being the subject of our study. Crystal structure data derive from various chemical and physical individuals and thus also represent sets of "impure" information. The aim of statistical endeavour then is to obtain the purest possible information using some suitable statistical model. It seems that the connection between methods of modern statistical research and chemical intuition can lead to the solution of many a problem of structure and bonding in condensed matter. Statistical methods applied to sets of crystal structure data can be in this study useful for a) objectification of conclusions and b) the creation of statistical models for existing facts. Such a phenomenological approach connected with chemical systematism of components of matter can lead to classification of individuals (crystal structures) being one of the problems of *crystal chemistry*. The study of correlations of these phenomenological knowledges with electronic structure of entities forming the condensed matter is the aim of the contemporary *chemical physics*.

In this work we shall study predominantly coordination compounds of transition elements (with d orbitals) and incidentally also mention some central entities of f systems.

Acknowledgements

The authors are thankful to Prof. I. D. Brown from the Institute for Materials Research, McMaster University, Hamilton, Canada for stimulative discussions. We thank all, who gave permission to present figures already published.

Bratislava, Prague
March 1988

Fedor Valach
Jan Ondráček
Milan Melník

Contents

1 Statistics of Interatomic Distances

An ideal crystal is usually understood to be a regular array of ions or atoms. However, if we speak of positions of these particles with respect to their crystallographic origin, physical reality forces us to make some approximations. An experiment based on the diffraction of X-ray or neutron radiation by crystal makes it possible to obtain information on the positions of atoms or ions in the form of coordinates of local maxima of electron densities. General expression of this electron density is

$$\varrho(x, y, z) = \frac{1}{V} \sum_h \sum_k \sum_l F(hkl)\, e^{-2\pi i(hx+ky+lz)}$$

where x, y, z are the fractional coordinates of any point in the unit cell of the volume V. $F(hkl)$ is structure factor which belongs to reflexion indexed as hkl. It is a common complex number which may be expressed as

$$F(hkl) = |F(hkl)|\, e^{i\varphi(hkl)}$$

Values $|F(hkl)|$ are observed experimentally. Phases of the structure factors $\varphi(hkl)$ are results from X-ray analysis. In this way obtained image of the crystal is also a certain averaged picture of the real crystal consisting of deviations from regularity of the arrangement of structure particles in the space. The atom or ion position is thus given by coordinates in the crystallographic coordinate system which was obtained using the diffraction experiment by means of X-ray or neutron structure analysis of the monocrystal.

The discussed crystal consists of discrete particles. From a physical point of view it represents a homogeneous discontinuum [1]. In such a system generally an infinite quantity of distances between the structure particles can be calculated. Let us call them *interatomic distances*. On the contrary, however, in crystal structure these connecting lines define special relations at least between the systems which consist of the nucleus of charge $+Ze$ and $K = Z - z$ electrons, called by Jørgensen [2] *monatomic entities* and marked by symbol M^{z+} [2]. For the investigation of these relations by statistical approach it is necessary to randomly sample a set of substances with crystal structures containing monatomic entities of certain common proerties. It is then possible by calculation to obtain the final set of distances between the entities of certain types (e.g. of chemical elements of formal valency) contained in one crystallochemical unit. Based on group-symmetrical properties of the discontinuum it is possible to define particle groups in every crystal [3]. The set of particles belongs to the same *particle group G* only when every set in the crystal being structurally equivalent with it has in common with group G either all particles or none. To every particle group may be assigned the symmetry- and translation group of the discontinuum. These groups consist of translation and symmetry operations contained in some of 230 space groups. They may be divided into classes marked as $\Sigma_0 \Sigma_I \Sigma_{II}$ and Σ_{III} [3]. The index number of Σ gives the number of independent translations. According to this classification particle groups may be

(a) of threedimensionally finite geometry (Σ_0)
(b) of onedimensionally infinite geometry (Σ_I)
(c) of twodimensionally infinite geometry (Σ_{II})
(d) of threedimensionally infinite geometry (Σ_{III})

The above-mentioned particle groups have, however, only a formal meaning. If they have a crystal structural signification delimited they will at the same time, be building units of the structure. Figure 1.1 shows a building unit of the crystal structure of α-bis(1-carbamoyl-3,5-dimethylpyrazolato)copper(II) as also the packing of such units in the unit cell. Covalently bonded atoms form a slightly distorted, approximately planar building unit of finite geometry. In literature such a

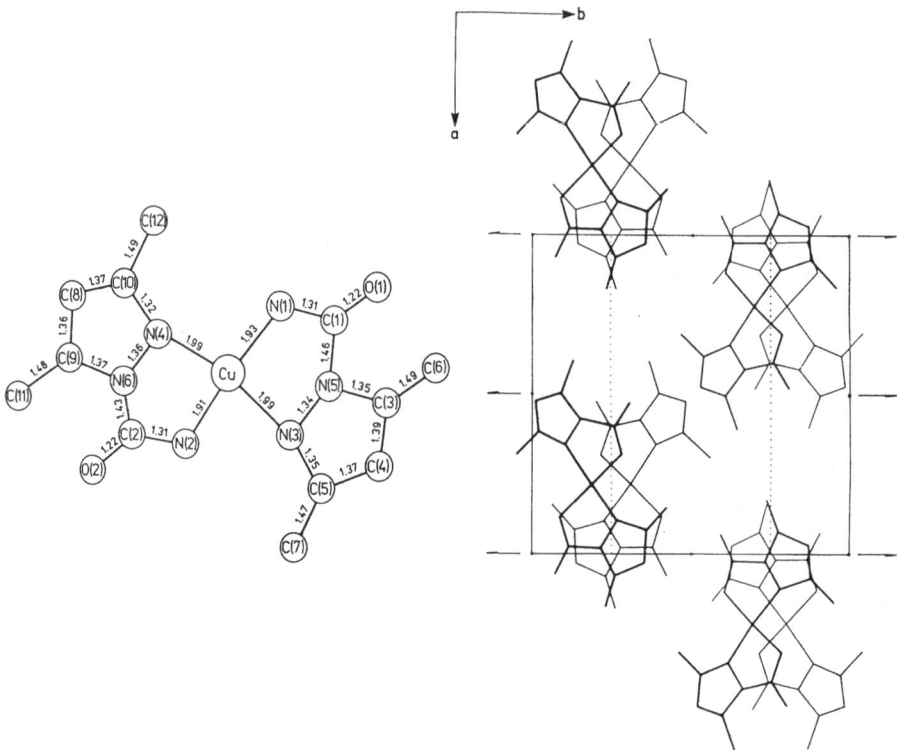

Fig. 1.1. Structure of the building unit of α-bis(1-carbamoyl-3,5-dimethylpyrazolato)copper(II) and arrangement of these units in the unit cell. Space group $P2_1/c$, monoclinic, $Z = 8$, $a = 13.70$, $b = 14.76$, $c = 7.46$ Å and $\beta = 114.01°$ [4]. Hydrogen atoms are omitted.

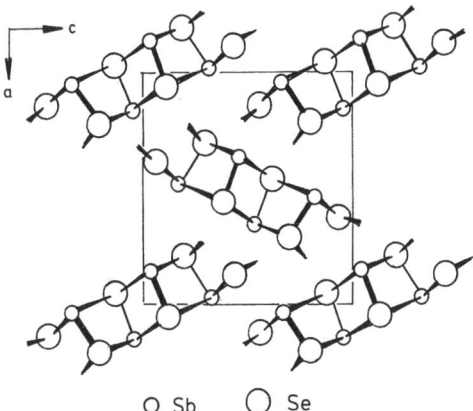

O Sb   O Se

Fig. 1.2. Chains of the crystal structure of Sb_2Se_3 [5] in the projection into plane (010). Orthorhombic, $Z = 4$, $a = 11.77$, $b = 3.962$ and $c = 11.62$ Å.

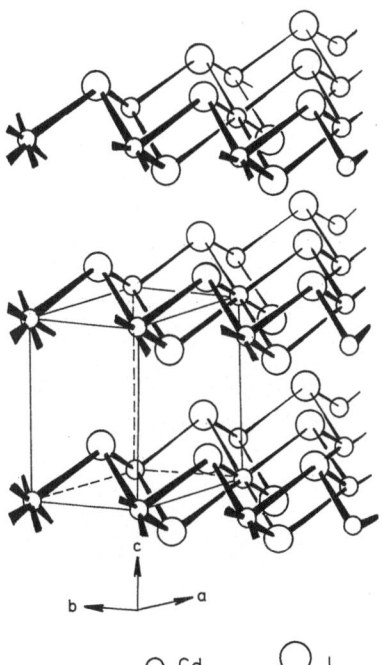

O Cd   O I

Fig. 1.3. Layers of the crystal structure of CdI_2 [6]. Hexagonal, $Z = 1$, $a = 3.543$ and $c = 5.039$ Å.

unit is also called *molecule* or *island*. The building units of Sb_2Se_3 structure form a *chain* (Fig. 1.2). The structure of CdI_2 (Fig. 1.3) consists of building units called *layers*. Structure of wurtzit (ZnS, Fig. 1.4) consists of the infinite building unit, called *skeleton*.

In order to delimitate building units in crystal structures it is necessary to introduce a criterion for the bonding of two monatomic entities. The nature of their mutual interaction, especially for entities of transition elements is, however, for many cases rather complicated. A more profound study of these interactions falls in the region of quantum chemistry and interpretation of spectral measurements of substances. Here

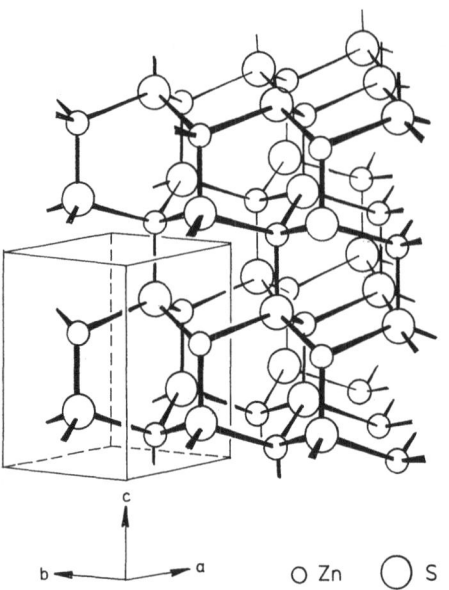

Fig. 1.4. Crystal structure skeleton of ZnS (wurtzit) [7]. Hexagonal, $Z = 2$, $a = 3.823$ and $b = 6.261$ Å. Full circles mark the unique set of monatomic entities.

○ Zn   ◯ S

we will limit ourselves to approximation using the conception of relative atomic sizes [8–26]. Figure 1.5 shows an example of interatomic potential. Irrespective of the interaction type, there always exists solely one equilibrium distance d_0. It is the distance corresponding to the equilibrium of repulsive and attractive forces. Near this energy minimum it is possible to introduce a criterion for the bonding covalency of two monatomic entities of radii r_A and r_B:

$$d \leqq r_A + r_B + T \tag{1.1}$$

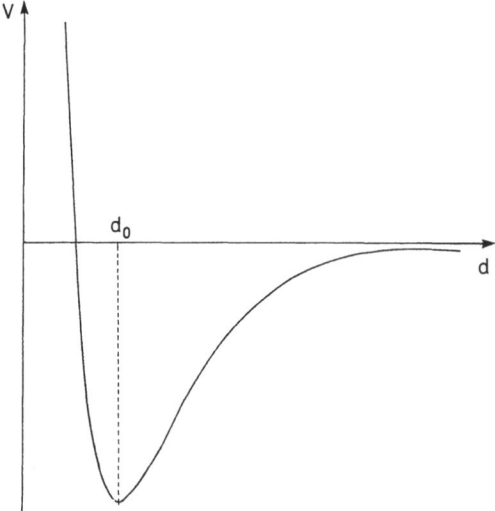

Fig. 1.5. Example of interatomic potential.

T is the elected tolerance. The distance between the entities d, corresponding to the inequality (1.1) will be called *bond length*. *Cambridge data retrieval package* [27] gives the possibility to obtain sets of such bond lengths for every accepted crystal structure containing at least one monatomic entity of organic carbon. This *Database* covers information on such structures published from 1935. It is used in connection with the bibliographic file (BIB) and connectivity (CON). These files decide which data sets from file DAT will be retrieved. Files BIB, CON and DAT are discussed in detail in [28–32]. We shall direct our attention to coordination compounds with atoms of transition elements. For every structure one can obtain atom coordinates for a complete *crystallochemical unit*. This unit consists of a unique set from discrete net in which monatomic entities are mutually bonded. The bonding is tested by criterion (1.1) with tolerance $T = 0.40\,\text{Å}$ ($1\,\text{Å} = 100\,\text{pm} = 0.1\,\text{nm}$). The radii used are listed in Table 1.1.

Table 1.1. Atomic radii used for the bonding criterion of monatomic entities in *Cambridge data retrieval package* [27] in Å

Ac	1.88	Eu	1.99	Nd	1.81	Se	1.22
Ag	1.59	F	0.64	Ni	1.50	Si	1.20
Al	1.35	Fe	1.34	Np	1.55	Sm	1.80
Am	1.51	Ga	1.22	O	0.68	Sn	1.46
As	1.21	Gd	1.79	Os	1.37	Sr	1.12
Au	1.50	Ge	1.17	P	1.05	Ta	1.43
B	0.83	Hf	1.57	Pa	1.61	Tb	1.76
Ba	1.34	Hg	1.70	Pb	1.54	Tc	1.35
Be	0.35	Ho	1.74	Pd	1.50	Te	1.47
Bi	1.54	I	1.40	Pm	1.80	Th	1.79
Br	1.21	In	1.63	Po	1.68	Ti	1.47
C	0.68	Ir	1.32	Pr	1.82	Tl	1.55
Ca	0.99	K	1.33	Pt	1.50	Tm	1.72
Cd	1.69	La	1.87	Pu	1.53	U	1.58
Ce	1.83	Li	0.68	Ra	1.90	V	1.33
Cl	0.99	Lu	1.72	Rb	1.47	W	1.37
Co	1.33	Mg	1.10	Re	1.35	Y	1.78
Cr	1.35	Mn	1.35	Rh	1.45	Yb	1.94
Cs	1.67	Mo	1.47	Ru	1.40	Zn	1.45
Cu	1.52	N	0.68	S	1.02	Zr	1.56
Dy	1.75	Na	0.97	Sb	1.46		
Er	1.73	Nb	1.48	Sc	1.44		

1.1 Some Statistical Concepts and Inferences

In order to elucidate and develop further techniques serving the study of properties of monatomic entities in condensed matter we will present some principles of statistics[1]. In first approximation we can say that the matter will be the endeavour to compile data based on which inferences could be drawn and decisions made. As it was already mentioned the crystallographic database can afford us a list of interatomic distances between monatomic entities of a certain type contained in the

---

[1] For further detailed explanations see [33–42].

unique set of crystallochemical unit of the crystal structure under consideration. Let us mark this list as $x_1, x_2, x_3, \ldots$. It is clear that in the case of a set of structures of substances of the studied chemical class (coordination compounds, with a certain type of central atom) we can obtain n such lists. The choice of any set may be considered an experiment. Since we randomly elected one from several cases, it will be a *random experiment*. To each list we can assign point A_i of a discrete sample space S. To each point of space S can further be assigned the weight $w_i \geq 0$, so that the expression will hold $\sum_{i=1}^{n} w_i = 1$. Any subset A of the sample space has then the probability $\Pr(A)$ equal to the sum of weights of the points involved in set A. The point A_i is also called *event* and the probability introduced this way is the *mathematical definition of probability*[2].

For the union of any two events $A_1 \cup A_2$ it holds

$$\Pr(A_1 \cup A_2) = \Pr(A_1) + \Pr(A_2) - \Pr(A_1 \cap A_2)$$

where $A_1 \cap A_2$ means their intersection. If events A_1, A_2 are mutually exclusive $(A_1 \cap A_2 = 0)$ then we have

$$\Pr(A_1 \cup A_2) = \Pr(A_1) + \Pr(A_2)$$

This result can be generalized for the case of n mutually exclusive events:

$$\Pr(A_1 \cup A_2 \cup A_3 \cup \ldots \cup A_n) = \Pr(A_1) + \Pr(A_2) + \Pr(A_3) + \ldots + \Pr(A_n)$$

When event A occurs under the condition that $\bar{A}$ does not occur and vice versa, i.e. they are mutually *complementary*, then for their probability it holds

$$\Pr(A) = 1 - \Pr(\bar{A})$$

The probability of event A_1, under the condition of an occurrence of event A_2 $(\Pr(A_2) \neq 0)$ is defined by the formula

$$\Pr(A_1 \mid A_2) = \frac{\Pr(A_1 \cap A_2)}{\Pr(A_2)}$$

When A_1 and A_2 are unreleated events, the probability is $\Pr(A_1 \mid A_2) = \Pr(A_1)$ and from the above formula we obtain

$$\Pr(A_1 \cap A_2) = \Pr(A_1) \Pr(A_2)$$

Again, this inference may be extended for the case of n mutually unrelated events:

$$\Pr(A_1 \cap A_2 \cap A_3 \cap \ldots \cap A_n) = \Pr(A_1) \Pr(A_2) \Pr(A_3) \ldots \Pr(A_n)$$

---

[2] For further definitions of probability see [39–42].

If the number of points of space S is infinitely great, then there exists an interval of values, which the crystal structure parameters can under consideration acquire. By this we introduced the *continuous random variable*. When this variable can acquire a final number of values, then it is a *discrete random variable*.

A continuous random variable ξ can occur in the interval of $-\infty < \xi \leq x$ with a probability, which we will sign as $\Pr (\xi \leq x)$. Then the function $P(x) = \Pr (\xi \leq x)$ will be called *probability distribution* or *distribution function* of the variable ξ. The function of $P(x)$ may be expressed by the formula

$$P(x) = \int_{-\infty}^{x} f(\xi) \, d\xi$$

The function of $f(\xi)$ is called the *density function* or *distribution density* of variable ξ. Thus this function is very important for the calculation of the probability of random variable. It enables us to introduce still further quantities characterizing the continuous random variable. It is the *expected value*

$$\mu = E(\xi) = \int_{-\infty}^{\infty} x f(x) \, dx \tag{1.1.1}$$

If ξ and η are two unrelated random variables and c_1 and c_2 are constants, then the above definition allows to deduce the rule:

$$E(c_1 \xi + c_2 \eta) = c_1 E(\xi) + c_1 E(\eta) \tag{1.1.2}$$

The *variance* of the random variable ξ is defined by the formula

$$\sigma^2 = \text{Var}(\xi) = E[(\xi - \mu)^2] = \int_{-\infty}^{\infty} (x - \mu)^2 f(x) \, dx \tag{1.1.3}$$

The quantity of $\sigma = \sqrt{\text{Var}(\xi)}$ is called the *standard deviation*. In the case of discrete random variable which can attain values x_i with probabilities p_i $(i = 1, 2, 3, ...)$, the expected value and variance are

$$\mu = \sum_{i=1}^{\infty} x_i p_i \qquad \sigma^2 = \sum_{i=1}^{\infty} (x_i - \mu)^2 \, p_i$$

From the above given definitions of the expected value and the variance of random variable it is easy to deduce the relation:

$$\sigma^2 = \text{Var}(\xi) = E[(\xi - \mu)^2] = E(\xi^2) - E(\xi)^2 = E(\xi^2) - \mu^2 \tag{1.1.4}$$

The distribution function as well as the density function are unknown in many statistical problems. The expected value and variance are, however, parameters characterizing these functions, or as we also say, they characterize the probability distribution. In order to describe it more completely, moments are used. The kth *moment* of the random variable is defined by the formula:

$$\mu_k = E(\xi^k) \qquad k = 1, 2, 3 \ldots$$

Similarly the kth *central moment* is

$$v_k = E[(\xi - \mu)^k]$$

From this point of view the expected value is the first moment, while the variance being the second central moment of random variable. The meaning of the variance is expressed by *Tchebischeff inequality*:

$$\Pr\left(|\xi - \mu| \geq \varepsilon\right) \leq \frac{\sigma^2}{\varepsilon^2}$$

for arbitrary $\varepsilon > 0$. Figure 1.1.1 shows the dependence of the maximum probability of acquiring the value of the variable outside the interval $\langle \mu - \varepsilon, \mu + \varepsilon \rangle$ from ε (full line). This probability vehemently falls with an increasing number of ε

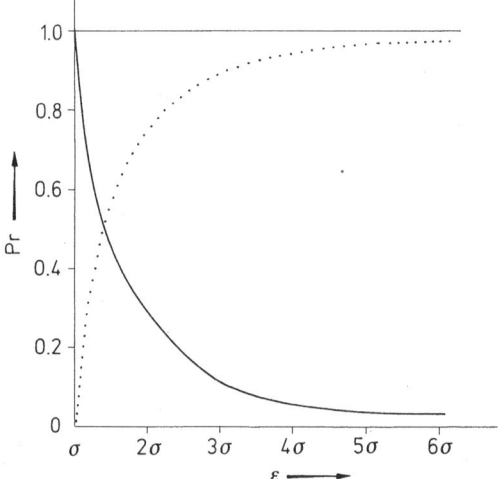

Fig. 1.1.1. Dependence of the maximum (full line) and the minimum (dotted line) probability from ε in Tschevbyscheff's inequality.

equalling a multiple of σ. The probability of the variable to acquire the value within the interval $(\mu - \varepsilon, \mu + \varepsilon)$ (being the complementary event to the former one), may be expressed by the inequality:

$$\Pr\left(|\xi - \mu| < \varepsilon\right) \geq 1 - \frac{\sigma^2}{\varepsilon^2}$$

The minimum value of this probability in dependence on ε is shown in Fig. 1.1.1 by the dotted line. Thus the standard deviation and the variance are the measure of spread of possible values of the random variable. Sometimes it appears advantageous for characterization of the relative spread to use standard deviation expressed in units of expected value:

$$v = \frac{\sigma}{\mu}$$

Parameter v is called the *variance coefficient*. For further characterization of probability distribution we shall also use the parameter called *asymmetry*, introduced by formula:

$$\gamma = \frac{v_3}{\sigma^3} = \frac{v_3}{[\mathrm{Var}\,(\xi)]^{3/2}} \tag{1.1.5}$$

The third central moment expressed by this way in units of standard deviation can exhibit a negative or positive value according to whether the tailing part of the distribution density is on the left side or on the right from the expected value (Fig. 1.1.2). Thus we speak of left-sided or right-sided asymmetry of distribution.

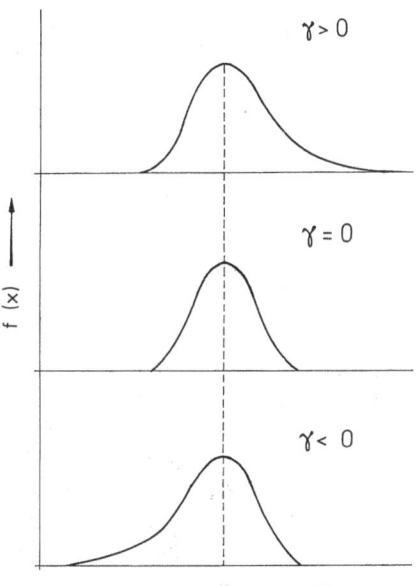

Fig. 1.1.2. Dependence of asymmetry γ on the shape of density function.

In many cases of solving statistical problems, however, we do not know the probability distribution of the random variable under study. Then it is necessary to form a presupposition of the distribution density $f(x)$. In this case we will have to create a *probability model*. The most used is a *normal probability model*. In this case we know the probability density:

$$f(x) = \frac{1}{\theta_2 \sqrt{2\pi}}\, e^{-(x-\theta_1)^2/2\theta_2} \qquad -\infty < x < \infty$$

Introducing this function into formulas (1.1.1) and (1.1.3) gives $\theta_1 = \mu$ and $\theta_2 = \sigma^2$ Fig. 1.1.3 respresents a graphical picture of this density. If $\mu = 0$ and $\sigma^2 = 1$, then variable exhibits *standard normal probability distribution*. Some other models of probability, which we shall use in various relations, are brought and graphically represented in Appendix A.

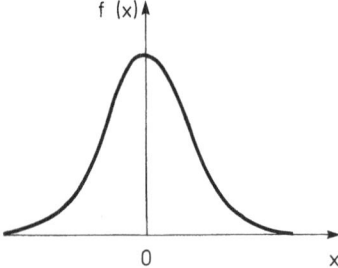

Fig. 1.1.3. Graph of the density of standard normal distribution.

In Table 1.1.1 their relations of μ, σ^2 and γ are listed. The dependence of the shape of density of normal distribution of the size of σ is shown in Fig. 1.1.4.

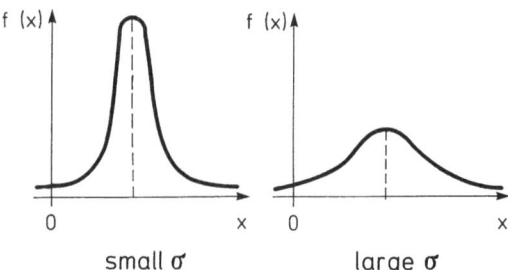

small σ large σ

Fig. 1.1.4. Dependence of the density of normal distribution on standard deviation σ.

In the actual case of the class of crystal structures under study we obtain the structure data of $x_1, x_2, x_3, \ldots, x_n$. This set may be considered as observed random sample of mutually unrelated random variables $X_1, X_2, X_3, \ldots, X_n$ with the same density of probability distribution. The characteristics of this distribution μ and σ^2 are not known. Therefore it is necessary to determine their best estimate, called the *point estimate*. This is why we shall introduce *sample characteristics* as *sample mean*

$$\bar{X} = \frac{\sum\limits_{i=1}^{n} X_i}{n} \quad \text{and } sample\ variance \quad S^2 = \frac{1}{n-1} \sum_{i=1}^{n} (X_i - \bar{X})^2$$

Let us calculate the expected value of the sample mean.

$$E(\bar{X}) = E\left(\frac{1}{n} \sum_{i=1}^{n} X_i\right) = \frac{1}{n} \sum_{i=1}^{n} E(X_i) = \mu$$

Table 1.1.1. Expected values, variances and asymmetries for some probability models (Appendix A)

Distribution	Expected value (μ)	Variance (σ^2)	Asymmetry (γ)
uniform	$(a + b)/2$	$(a - b)^2/12$	0
normal	μ	σ^2	0
Γ	τ/λ	τ/λ^2	$2/\sqrt{\tau}$
χ^2	ν	2ν	$2^{3/2}/\sqrt{\nu}$
exponential	$1/\lambda$	$1/\lambda^2$	2.0
β	$\dfrac{\tau}{\tau + \tau'}$	$\dfrac{\tau\tau'}{(\tau + \tau')(\tau + \tau' + 1)}$	a
Weibull	$\Gamma\left(\dfrac{1}{\tau} + 1\right)$	$\left\{\Gamma\left(\dfrac{2}{\tau} + 1\right) - \left[\Gamma\left(\dfrac{1}{\tau} + 1\right)\right]^2\right\}$	b
F	$\nu_2/(\nu_2 - 2)$ $\nu_2 > 2$	c	d
t	0	$\nu/(\nu - 2)$ $\nu > 2$	0

a $\quad \gamma = \dfrac{2(\tau - \tau')(\tau + \tau' + 1)^{1/2}}{(\tau\tau')^{1/2}(\tau + \tau' + 2)}$

b $\quad \gamma = \dfrac{\Gamma\left(1 + \dfrac{3}{\tau}\right) - 3\Gamma\left(1 + \dfrac{2}{\tau}\right)\Gamma\left(1 + \dfrac{1}{\tau}\right) + 2\left[\Gamma\left(1 + \dfrac{1}{\tau}\right)\right]^3}{\left\{\Gamma\left(1 + \dfrac{2}{\tau}\right) - \left[\Gamma\left(1 + \dfrac{1}{\tau}\right)\right]^2\right\}^{3/2}}$

c $\quad \sigma^2 = 2\nu_2^2(\nu_1 + \nu_2 - 2)/\nu_1(\nu_2 - 2)^2(\nu_2 - 4) \quad \nu_2 > 4$

d $\quad \gamma = \dfrac{(2\nu_1 + \nu_2 - 2)[8(\nu_2 - 4)]^{1/2}}{(\nu_2 - 6)(\nu_1 + \nu_2 - 2)^{1/2}} \quad \nu_2 > 6$

Because of equal probability distribution of the variables X_i is

$$\sum_{i=1}^{n} E(X_i) = nE(X_i) = n\mu$$

Similarly the following identity can be proved (Appendix B):

$$E(S^2) = \sigma^2$$

We say that sample mean and sample variance are *unbiased estimates* of the expected value and variance. For practical calculations of the *observed sample characteristics* the formulas are used

$$\bar{x} = \frac{\sum\limits_{i=1}^{n} x_i}{n}$$

$$s^2 = \frac{n \sum\limits_{i=1}^{n} x_i^2 - \left(\sum\limits_{i=1}^{n} x_i^2\right)^2}{n(n-1)}$$

In solving statistical problems we often have a set of data $x_1, x_2, x_3, \ldots, x_n$, which we assume to be an observed sample of random variables $X_1, X_2, X_3, \ldots, X_n$, to which the normal model of probability applies. We thus assume the set under study to be a random sample from normal distribution.

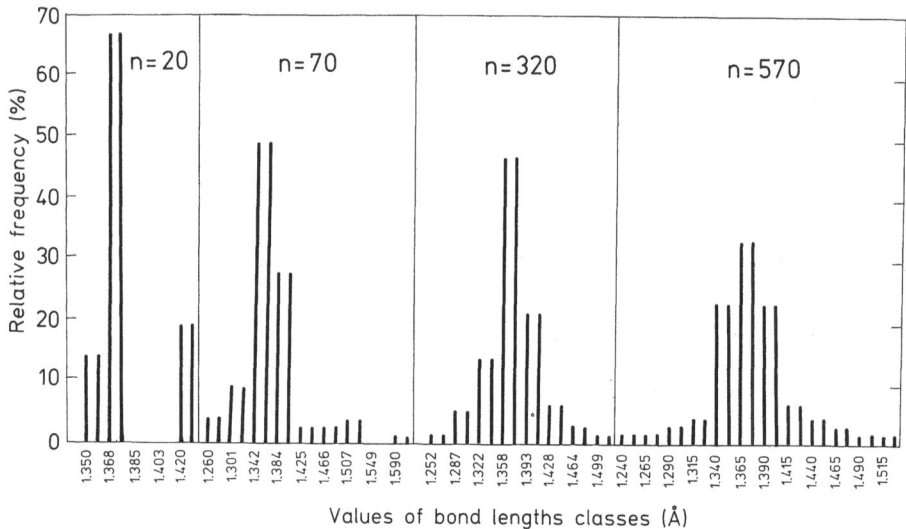

Fig. 1.1.5. Empirical distributions of bond lengths C-C in pyridine rings for some sample sizes.

Figure 1.1.5 represents a histogram (see Chap. 1.2) of bond lengths C—C in a pyridine molecule for a different number of sample bonds. As these histograms show, the distribution of relative frequencies approximates the distribution similar to normal (Fig. 1.1.4), the greater the sample size n is. This fact is expressed by the *central limit theorem* of mathematical statistics, according to which the random variable

$$Z = \frac{\bar{X} - \mu}{\sigma} \sqrt{n} \tag{1.1.6}$$

exhibits a distribution converging with increasing n to standard normal distribution.

One of the perhaps most significant statistical approaches for the solution of our problems is *testing statistical hypotheses*. A detailed explanation of this part of statistics is involved in [43]. The statistical hypothesis is understood to be a presumption required from the distribution of random variables under study and from the parameters of this distribution. Testing the hypothesis marked as H_0 means to verify the accuracy of the formulated hypothesis. Against the hypothesis H_0 (also called zero hypothesis) we put the alternative hypothesis H_1. Let us consider the case of a parameter θ of the random variable X distribution. Then we can formulate hypothesis H_0 as $\theta = \theta_0$, where θ_0 is a certain value of parameter θ against the alternative hypothesis $H_1 : \theta \neq \theta_0$. In practical problems arises the question, which of these hypotheses should we accept and which is to be rejected. Hypothesis H_0 and alternative hypothesis H_1 may be expressed also as follows:

(a) $H_0 : \theta = \theta_0$ against $H_1 : \theta < \theta_0$
(b) $H_0 : \theta = \theta_0$ against $H_1 : \theta > \theta_0$

These cases are one-sided alternative hypotheses. The choice of hypothesis and alternative hypothesis depends on the nature of the statistical problem. Such a decision-making process can be, however, performed with a certain inaccuracy. The fault arising by an incorrect rejection of the hypothesis H_0 is called *fault of the 1st type*. The fault arising by incorrect rejection of the alternative hypothesis H_1 is the *fault of the 2nd type*. The probability of the 1st type fault marked with α, gives the *significance level*, while the probability of the 2nd type fault β is the *test power* of hypothesis H_0. It expresses the probability of its rejection. Let us take sample characteristic $h(x)$. Then the probability of the 1st type fault may be written as

$$\Pr\,(h(x) \in M_\alpha) = \alpha$$

Thus the probability of occurrence of characteristic $h(x)$ in the set M_α called the *critical region* is equal to the significance level. Then the probability of the 2nd type fault can be written as

$$\Pr\,(h(x) \in M_\beta) = \beta$$

or

$$\Pr\,(h(x) \notin M_\alpha) = 1 - \Pr\,(h(x) \in M_\alpha) = 1 - \beta$$

Figure 1.1.6 represents the critical region of the random variable with normal and F distribution. In practical problems usually a significance level α is selected and a suitable test (critical area M_α) is elected minimizing the 2nd type fault, i.e. it maximizes the test power β. Such a test is called the strongest one.

Using tests of statistical hypotheses enables us to make rapid decisions, above all of parameters of random variables with normal distribution. For these purposes a number of standard procedures have been developed. Usually they consist of the following steps:

(a) choice of the significance level α
(b) formulation of the hypothesis H_0

(c) choice of the testing characteristics or the criterion
(d) interpretation of the decision result.

For our purposes we will direct our attention in more detail to *sample size problem*. Using the central limit theorem the characteristics (1.1.6) exhibit a distribution near the standard normal one. The assumption of a normal distribution of

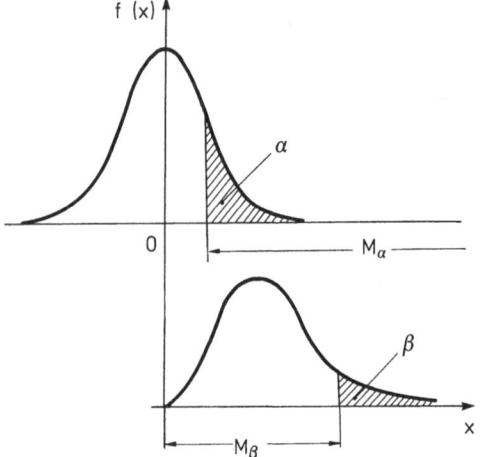

Fig. 1.1.6. Example of a critical region for normal and *F* distribution.

variable Z is approximately correct, providing that n is not too small. Therefore also testing the sample mean value with sufficient accuracy does not require a rigorous presupposition of a normal distribution of the studied random variable. Besides point estimates of distribution parameters, interval estimates can also be tested. For the variable Z a hypothesis can be formulated as:

$$H_0:|Z| \geq z'_\alpha \quad \text{against} \quad H_1:|Z| < z'_\alpha$$

where z'_α is the *critical value* of variable $|Z|$. For practical purposes, however, it is advantageous to formulate a twosided hypothesis:

$$H_0: Z \geq z_{\alpha/2} \quad \text{against} \quad H_1: Z < z_{\alpha/2}$$

and

$$Z \leq z_{\alpha/2} \quad \text{against} \quad Z > z_{\alpha/2}$$

The critical areas for these hypotheses are shown in Fig. 1.1.7. For an interval outside the critical area we may write:

$$\Pr\left(-z_{1-\alpha/2} < Z < z_{1-\alpha/2}\right) = 1 - \alpha$$

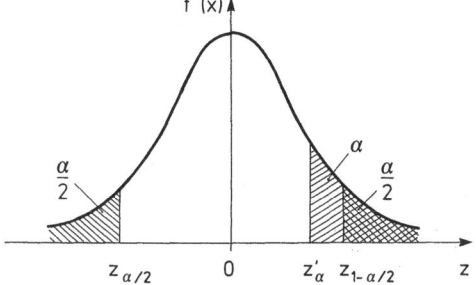

Fig. 1.1.7. Critical regions of variable Z of one-sided and two-sided hypotheses.

and transform it to inequality

$$\Pr\left(\bar{X} - z_{1-\alpha/2}\,\frac{\sigma}{\sqrt{n}} < \mu < \bar{X} + z_{1-\alpha/2}\,\frac{\sigma}{\sqrt{n}}\right) = 1 - \alpha \qquad (1.1.7)$$

In this way limited random interval for parameter μ is called the *confidence interval* and the probability $1 - \alpha$ is the *confidence coefficient*. Thus the confidence interval involves parameter μ with the probability $1 - \alpha$. The coefficient $z_{1-\alpha/2}$ is tabulated for various probability values [44].

The length of the confidence interval is

$$\bar{X} + z_{1-\alpha/2}\,\frac{\sigma}{\sqrt{n}} - \left(\bar{X} - z_{1-\alpha/2}\,\frac{\sigma}{\sqrt{n}}\right) = 2z_{1-\alpha/2}\,\frac{\sigma}{\sqrt{n}}$$

We have to find out how n should be in order that the length of this interval does not exceed the width of the class of one data group, which we will mark as Δ:

$$2z_{1-\alpha/2}\,\frac{\sigma}{\sqrt{n}} < \Delta$$

From this it follows

$$n > \left(\frac{2\sigma z_{1-\alpha/2}}{\Delta}\right)^2$$

The *critical sample size* is

$$n_{\text{crit}} = \left(\frac{2\sigma z_{1-\alpha/2}}{\Delta}\right)^2 \qquad (1.1.8)$$

Solving the sample size problem using this inequality presupposes, however, knowledge of the standard deviation σ. Since this parameter is usually not known, it is more advantageous to use the characteristics:

$$T_{n-1} = \frac{\bar{X} - \mu}{S}\sqrt{n}$$

The variable T_{n-1} exhibits Student t distribution (Appendix A) with $n-1$ degrees of freedom. Here a random character of sample of $X_1, X_2, X_3, \ldots, X_n$ is assumed from the normal distribution with expected value μ. Empirical experience shows, however, that variable T_{n-1} has distribution t, though the normality condition is not kept providing that n is not too small. Applying the inequality (1.1.7) to variable T_{n-1} we obtain the formula

$$n_{crit} = \left(\frac{2st_{n-1;1-\alpha/2}}{\varDelta}\right)^2 \tag{1.1.9}$$

Tables of the critical values of $t_{n-1;1-\alpha/2}$ are presented in [44].

In the case of two random variables ξ and η *covariance* is introduced by the formula:

$$\mathrm{Cov}\,(\xi, \eta) = E[(\xi - E(\xi))\,(\eta - E(\eta))]$$

The *correlation coefficient* being the measure of the mutual dependence of these variables is

$$\varrho_{\xi,\eta} = \frac{\mathrm{Cov}\,(\xi, \eta)}{\sqrt{\mathrm{Var}\,(\xi)\,\mathrm{Var}\,(\eta)}} \tag{1.1.10}$$

If continuous random variables ξ and η are independent ($\varrho_{\xi,\eta} = 0$), then the probability density of the *random vector* $\omega = (\xi, \eta)$ is

$$f_\omega(x_1, x_2) = f_\xi(x_1)\,f_\eta(x_2)$$

where f_ξ and f_η are the densities of ξ and η variables.

In the case of two random samples from two sets with normal distributions of equal variances, the ratio of their sample characteristics $F = S_1^2/S_2^2$ exhibits Fischer — Snedecor distribution (F distribution in Appendix A).

If $X_1, X_2, X_3, \ldots, X_n$ are random variables, each with standard normal distribution, then variable $X = \sum\limits_{i=1}^{n} X_i^2$ shows distribution χ_n^2 (Appendix A). This statement can also be extended to arbitrary quadratic forms[1] of variables X_i, for which it holds $Q_1 + Q_2 + Q_3 + \ldots + Q_u = \sum\limits_{i=1}^{n} X_i^2$. According to the Cochran theorem [36] these quadratic forms are independent and each of them has the distribution $\chi_{n_i}^2$ (n_i is the rank of form Q_i) only when $\sum\limits_{i=1}^{u} n_i = n$

---

[1] The quadratic form of rank n_i of variables $X_1, X_2, X_3, \ldots, X_n$ is defined by the formula

$$Q_{n_i} = \sum\limits_{j=1}^{n_i} \sum\limits_{i=1}^{n_i} a_{ij}X_iX_j \quad \text{where } a_{ij} \text{ are real constants.}$$

1.2 Statistics of Lengths of Interatomic Vectors

Statistical procedures can be used for the purpose of an objective analysis of great sets of data describing molecular stereochemistry. This can be the already mentioned bond lengths of a certain type, or structure parameters sampled according to the nature of the given problem. So in the case of construction of techniques for classification and quantification of differences in the geometry of the set of 48 tripeptide fragments forming β-loops, Murray-Rust and Raftery [45] applied atomic coordinates in the intern Carthesian system, its origin being in the molecule centroid. Doms *et al.* [46] used for their statistical studies of norbornane fragments, sets of bond lengths and bond and torsion angles.

The geometry of crystal structure can be described in the three dimensional Euclidian space, also called *point space*. The structure can, however, also be described by introducing the *vector space*. According to *International Tables of Crystallography* [47] the vector space is defined as follows:

(a) Assigned to arbitrary points P and Q of n-dimensional point space E^n is vector $\overrightarrow{PQ} = r$ of the vector space V^n.
(b) To each point P of the point space E^n and to each vector r of the space V^n there exists only one point Q, of space E^n, for which it holds $\overrightarrow{PQ} = r$
(c) $\overrightarrow{PQ} + \overrightarrow{QR} = \overrightarrow{PR}$

The set of all the vectors mutually connecting the monatomic entities of the crystal structure, which we shall call *interatomic vectors*, is part of the vector space. If in the point space arrays occur covering the whole space and if any of them have no common inner point, then we say that there is a *space division*. Such a division, for which a group of operations exist transforming it into itself, is called a *regular division*. For two arbitrary arrays of regular division there exists at least one operation from group G mutually transforming them into themselves. A regular division of a plane consists of planiogons, while the division of a three-dimensional point space consists of stereohedrons. The regular division is *normal*, when its arrays are limited by convex polyhedrons being mutually boarded by common faces [48]. For each convex polyhedron of the normal division of a point space Euler formula holds:

$$f + v = e + 2$$

f being the number of faces, v being the number of vertices while e gives the number of edges.

Each lattice point in the point space can be expressed as the terminal point of the vector

$$r_{u,v,w} = u\boldsymbol{a} + v\boldsymbol{b} + w\boldsymbol{c} \tag{1.2.1}$$

where $\boldsymbol{a}$, $\boldsymbol{b}$, $\boldsymbol{c}$ form the basis of the lattice and u, v, w are integers. In order to describe the vector space properties, it is useful to introduce the *domain of influence* [49], sometimes also called *Dirichlet domain, Voronoi domain*, or *Wirkungsbereich* [50]. This is an array consisting of all the points of the space being nearer to the lattice point under consideration, than to any other lattice point. The domain of influence of the lattice point **A** is then limited by a convex polyhedron, being the result of the following construction:

a) connect point **A** with all other points of the lattice
b) divide these connecting lines by bisectral planes
c) the innermost polyhedron limited by these planes is the domain of influence of point **A**.

When the arrays of regular space division are domains of influence, then this division is normal [51]. Minkowski [52] proved that in the point space E^n the minimum and maximum number of faces of the polyhedron constructed in this way is $2n$ and $2(2^n - 1)$. Polyhedrons of this type in a threedimensional Euclidian space ($n = 3$) are described and tabulated e.g. in [53–55]. Point **A** is the *geometrical centre* of its influence domain. Let us place it into the position of the central monatomic entity of a concrete crystal structure. Let the remaining monatomic entities be connected with point **A** by interatomic vectors. The common origin of these vectors is itself in point A (Fig. 1.2.1). For each other point of the space, being a translational image of point **A**, such a set of vectors is congruent. Delone et al. [56] proved,

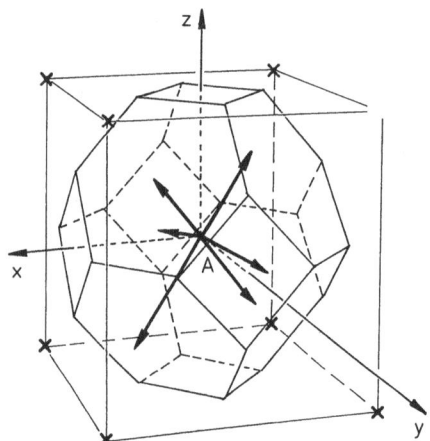

Fig. 1.2.1. Domain of influence of point **A** of cubic lattice with 1-centered unit cell. Interatomic vectors with common origin in point **A** and with terminal points contained in the influence domain.

independently of group theoretical properties of the discontinuum that the congruence of such "spiders" is the criterion for the regularity of the crystal structure. The infinite set of interatomic vectors, connecting point **A** with the positions of other entities of the crystal structure (with common origin in point **A**) is an unambiguously determined subset of vectors, the terminal points of which are involved in the domain of influence of point **A** (Fig. 1.2.2). The topological properties of the central monatomic entity can thus be studied within the influence domain of its *site* (see Chap. 3). Such properties of monatomic entities of crystal structures are expressed by the *Vector Equilibrium Principle (VEP)*, as formulated by Loeb [57]:

Crystal structures tend to assume configurations, in which a maximum number of identical monatomic entities are equidistant from each other; if more than a single

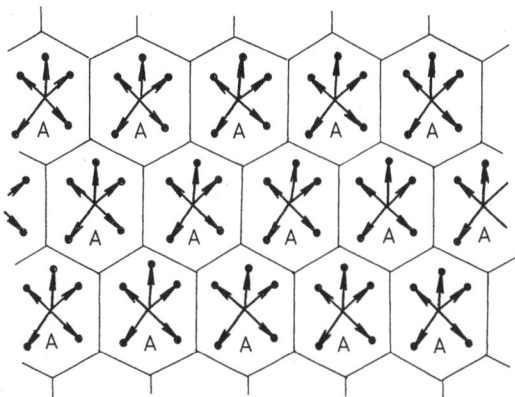

Fig. 1.2.2. The subsets of interatomic vectors connecting point **A** with the positions of monatomic entities, contained in its influence domain. This subset with its translation images describes the positions of all entities in the crystal structure. A two-dimensional case.

type of monatomic entities are present, each monatomic entity tends to be equidistant from as many as possible of each type of entities.

The tendencies formulated in *VEP* are the consequences of the equilibrium of attractive and repulsive forces between monatomic entities in the structure. Laves [58] postulated similar principles based on a classification of intermetallics:

(a) *Space principle*
(b) *Symmetry principle*
(c) *Connection principle*

Principle (*a*) expresses the tendency of monatomic entities of the shape of a rigid sphere to fill in the space as effectively as possible. Principles (*b*) and (*c*) express the tendency of entities to occupy positions in the structure with the greatest possible symmetry, and to form the greatest quantity in the shortest interatomic distances. If all monatomic entities would ideally fulfil the above principles, the substances should crystallize in cubic, closely packed structures (Fig. 1.2.3). In structures of inorganic compounds, however, there are often hindrances as a consequence e.g. the bond factor

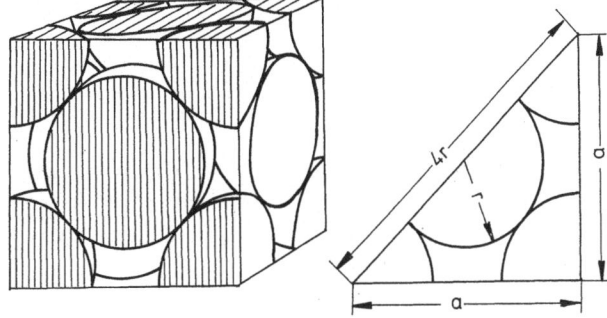

Fig. 1.2.3. Cubic close packed crystal structure. F-centered cubic unit cell (fcc structure).

and the needs of stoichiometry. *VEP* coincides with principles (a)–(c). The first of them, however, is modified on the condition that it does not presuppose monatomic entities in the form of rigid spheres, but it also admits certain soft interatomic contacts, i.e. the *non-rigidity* of monatomic entities. The cubic structure types as classified by Loeb [57] satisfy *VEP* most truly. These structures may be described by means of Pearson's classification scheme [59] based on fourteen Bravais lattices and the number of atoms per unit cell. In the case of coordination compounds of transition, element atom deviations from *VEP* may be expected, caused by the complicated nature of bond interactions of the central monatomic entity with ligands. Statistical study of these bond lengths [60] shows, however, a partial contribution of central monatomic entities of this type to the vector equilibrium of the crystal, also in their inner coordination spheres. Usually this contribution is mostly shown when a central entity is contained in a threedimensionally relatively isolated, final structure unit. However, if we consider lengths of interatomic vectors contained in the influence domain of the position of central monatomic entity of more distant entities, there are often interactions of electrostatistic and/or Van der Waals type. Such interactions between monatomic entities usually show an interaction potential close to the potential presented in Fig. 1.5. The set of entities with such interactions in agreement with the principle of the minimum of thermodynamic potential $F = U - TS + pV$ will best satisfy the principles formulated by Laves and *VEP*. The greatest deviations from the tendencies contained in these principles may also be expected in cases of greater distances between entities, when they are contained in a structure unit of mutually covalently bonded networks of the classes Σ_I, Σ_{II} and Σ_{III}. In our case of coordination compounds of transition elements, these networks are for the most part of the type Σ_0.

For statistical study of properties of the central monatomic entity of a certain type, we will use lengths of interatomic vectors, with a common origin in its site and the terminal points within its domain of influence. The so obtained sets of data using *Cambridge Crystallographic Database* are of great size and represent a numerical material, the extent of which is unclear. Therefore it is necessary to make a suitable adjustment of the obtained numerical sets. For our purposes it will be the most suitable way to group these numerical data into *classes*. The data of every set can be adjusted into succession:

$$x_1 \leqq x_2 \leqq x_3 \leqq \dots \leqq x_n$$

Each value will then be assigned into i^{th} class, when the inequality

$$h_i - \frac{\Delta}{2} \leqq x_j < h_i + \frac{\Delta}{2}$$

is fulfilled. h_i is the *middle of the class* and Δ is the *length of the class*. In this way, performed grouping of data presupposes equal length Δ for each class. Its size depends on the accuracy of the experimental method, i.e. of the monocrystal X-ray or neutron structure analysis as on the sample size. Experience shows that using a length of $\Delta = 0.05$ Å for each class is the most suitable. The convenience of choosing a class of this breadth for the given sample size can be tested by the criterion (1.1.9). When

f_i is the *frequence of occurrence* of interatomic vector lengths values of ith class of middle h_i, then it holds

$$\sum_{i=1}^{l} f_i = n$$

The *relative frequency* of ith class is f_i/n. Then the grouped data can be used for descriptive purposes. In Table 1.2.1 the grouped values of monatomic vector lengths

Table 1.2.1. Frequencies of the lengths of interatomic vectors with common origin in the site of central monatomic entity for coordination compounds of Ni^{2+} The terminal point of each vector is contained in the influence domain of the Ni^{2+} site.

Middle of class h_i [Å]	Frequence f_i	Relative frequence f_i/n [%]	Middle of class h_i [Å]	Frequence f_i	Relative frequence f_i/n [%]
0.575	2	0.0	2.425	1096	0.91
0.625	0	0.0	2.475	642	0.53
0.675	5	0.0	2.525	174	0.14
0.725	0	0.0	2.575	237	0.20
0.775	1	0.0	2.625	193	0.16
0.825	10	0.01	2.675	864	0.72
0.875	4	0.0	2.725	387	0,32
0.925	6	0.0	2.775	576	0.48
0.975	11	0.01	2.825	1119	0.93
1.025	25	0.02	2.875	1397	1.16
1.075	8	0.01	2.925	1911	1.59
1.125	16	0.01	2.975	1267	1.05
1.175	24	0.02	3.025	1952	1.62
1.225	7	0.01	3.075	1545	1.29
1.275	23	0.02	3.125	994	0.83
1.325	26	0.02	3.175	1037	0.86
1.375	28	0.02	3.225	802	0.67
1.425	22	0.02	3.275	649	0.54
1.475	37	0.03	3.325	733	0.61
1.525	26	0.02	3.375	799	0.66
1.575	18	0.02	3.425	627	0.52
1.625	21	0.02	3.475	662	0.55
1.675	21	0.02	3.525	618	0.51
1.725	42	0.04	3.575	477	0.40
1.775	96	0.08	3.625	537	0.45
1.825	334	0.28	3.675	586	0.49
1.875	2282	1.90	3.725	636	0.53
1.925	525	0.44	3.775	757	0.63
1.975	580	0.48	3.825	520	0.43
2.025	1182	0.98	3.875	531	0.44
2.075	1863	1.55	3.925	514	0.43
2.125	1704	1.42	3.975	626	0.52
2.175	1145	0.95	4.025	752	0.63
2.225	765	0.64	4.075	799	0.66
2.275	253	0.21	4.125	1109	0.92
2.325	269	0.22	4.175	1395	1.16
2.375	574	0.48	4.225	1414	1.18

Table 1.2.1 (continued)

Middle of class h_i [Å]	Frequence f_i	Relative frequence f_i/n [%]	Middle of class h_i [Å]	Frequence f_i	Relative frequence f_i/n [%]
4.275	1094	0.91	6.825	643	0.53
4.325	1306	1.09	6.875	725	0.60
4.375	1302	1.08	6.925	625	0.52
4.425	926	0.77	6.975	555	0.46
4.475	1131	0.94	7.025	464	0.39
4.525	692	0.58	7.075	632	0.53
4.575	798	0.66	7.125	492	0.41
4.625	747	0.62	7.175	514	0.43
4.675	1074	0.89	7.225	533	0.44
4.725	1108	0.92	7.275	760	0.63
4.775	955	0.79	7.325	424	0.35
4.825	843	0.70	7.375	468	0.39
4.875	759	0.63	7.425	504	0.42
4.925	877	0.73	7.475	499	0.42
4.975	905	0.75	7.525	478	0.40
5.025	905	0.75	7.575	438	0.36
5.075	898	0.75	7.625	360	0.30
5.125	831	0.69	7.675	353	0.29
5.175	746	0.62	7.725	434	0.36
5.225	871	0.73	7.775	373	0.31
5.275	1342	1.12	7.825	384	0.32
5.325	880	0.73	7.875	458	0.38
5.375	1003	0.83	7.925	347	0.29
5.425	1008	0.84	7.975	414	0.34
5.475	1182	0.98	8.025	368	0.31
5.525	1069	0.89	8.075	284	0.24
5.575	1104	0.92	8.125	265	0.22
5.625	856	0.71	8.175	410	0.34
5.675	772	0.64	8.225	342	0.28
5.725	773	0.64	8.275	243	0.20
5.775	913	0.76	8.325	273	0.23
5.825	1017	0.85	8.375	275	0.23
5.875	886	0.74	8.425	350	0.29
5.925	992	0.83	8.475	271	0.23
5.975	705	0.59	8.525	242	0.20
6.025	688	0.57	8.575	262	0.22
6.075	699	0.58	8.625	317	0.26
6.125	710	0.59	8.675	208	0.17
6.175	671	0.56	8.725	255	0.21
6.225	689	0.57	8.775	272	0.23
6.275	732	0.61	8.825	176	0.15
6.325	740	0.62	8.875	241	0.20
6.375	868	0.72	8.925	366	0.30
6.425	716	0.60	8.975	230	0.19
6.475	596	0.50	9.025	215	0.18
6.525	985	0.82	9.075	191	0.16
6.575	668	0.56	9.125	160	0.13
6.625	585	0.49	9.175	193	0.16
6.675	721	0.60	9.225	242	0.20
6.725	701	0.58	9.275	182	0.15
6.775	551	0.46	9.325	160	0,13

Middle of class h_i [Å]	Frequence f_i	Relative frequence f_i/n [%]	Middle of class h_i [Å]	Frequence f_i	Relative frequence f_i/n [%]
9.375	181	0.15	12.025	64	0.05
9.425	174	0.14	12.075	62	0.05
9.475	183	0.15	12.125	70	0.06
9.525	145	0.12	12.175	68	0.06
9.575	106	0.09	12.225	54	0.04
9.625	227	0.19	12.275	70	0.06
9.675	199	0.17	12.325	19	0.02
9.725	172	0.14	12.375	47	0.04
9.775	127	0.11	12.425	49	0.04
9.825	113	0.09	12.475	57	0.05
9.875	137	0.11	12.525	20	0.02
9.925	196	0.16	12.575	41	0.03
9.975	198	0.16	12.625	40	0.03
10.025	153	0.13	12.675	51	0.04
10.075	153	0.13	12.725	42	0.03
10.125	115	0.10	12.775	70	0.06
10.175	239	0.19	12.825	65	0.05
10.225	108	0.09	12.875	34	0.03
10.275	132	0.11	12.925	42	0.04
10.325	151	0.13	12.975	41	0.03
10.375	74	0.06	13.025	44	0.04
10.425	119	0.10	13.075	83	0.07
10.475	85	0.07	13.125	46	0.04
10.525	95	0.08	13.175	47	0.04
10.575	107	0.09	13.225	59	0.05
10.625	92	0.08	13.275	67	0.06
10.675	69	0.06	13.325	34	0.03
10.725	127	0.11	13.375	54	0.05
10.775	96	0.08	13.425	81	0.07
10.825	80	0.07	13.475	53	0.04
10.875	88	0.07	13.525	43	0.04
10.925	107	0.09	13.575	41	0.03
10.975	130	0.11	13.625	48	0.04
11.025	83	0.07	13.675	37	0.03
11.075	138	0.11	13.725	43	0.04
11.125	105	0.09	13.775	100	0.08
11.175	134	0.11	13.825	19	0.02
11.225	116	0.10	13.875	34	0.03
11.275	113	0.09	13.925	26	0.02
11.325	56	0.05	13.975	49	0.04
11.375	153	0.13	14.025	87	0.07
11.425	64	0.05	14.075	70	0.06
11.475	67	0.06	14.125	34	0.03
11.525	62	0.05	14.175	87	0.07
11.575	97	0.08	14.225	28	0.02
11.625	76	0.06	14.275	87	0.07
11.675	74	0.06	14.325	26	0.02
11.725	130	0.11	14.375	57	0.05
11.775	86	0.07	14.425	41	0.03
11.825	68	0.06	14.475	48	0.04
11.875	87	0.07	14.525	26	0.02
11.925	44	0.04	14.575	49	0.04
11.975	127	0.11	14.625	33	0.03

Table 1.2.1 (continued)

Middle of class h_i [Å]	Frequence f_i	Relative frequence f_i/n [%]	Middle of class h_i [Å]	Frequence f_i	Relative frequence f_i/n [%]
14.675	50	0.04	17.175	24	0.02
14.725	28	0.02	17.225	48	0.04
14.775	54	0.05	17.275	87	0.07
14.825	63	0.05	17.325	20	0.02
14.875	66	0.05	17.375	60	0.05
14.925	38	0.03	17.425	27	0.02
14.975	4	0.0	17.475	27	0.02
15.025	63	0.05	17.525	24	0.02
15.075	40	0.03	17.575	31	0.03
15.125	12	0.01	17.625	74	0.06
15.175	53	0.04	17.675	52	0.04
15.225	57	0.05	17.725	77	0.06
15.275	38	0.03	17.775	32	0.03
15.325	44	0.04	17.825	30	0.03
15.375	61	0.05	17.875	34	0.03
15.425	26	0.02	17.925	45	0.04
15.475	31	0.03	17.975	10	0.01
15.525	44	0.04	18.025	5	0.0
15.575	11	0.01	18.075	19	0.02
15.625	63	0.05	18.125	24	0.02
15.675	50	0.04	18.175	22	0.02
15.725	66	0.05	18.225	67	0.06
15.775	10	0.01	18.275	71	0.06
15.825	27	0.02	18.325	14	0.01
15.875	49	0.04	18.375	26	0.02
15.925	60	0.05	18.425	40	0.03
15.975	78	0.06	18.475	19	0.02
16.025	52	0.04	18.525	21	0.02
16.075	36	0.03	18.575	31	0.03
16.125	34	0.03	18.625	62	0.05
16.175	63	0.05	18.675	33	0.03
16.225	41	0.03	18.725	49	0.04
16.275	46	0.04	18.775	68	0.06
16.325	35	0.03	18.825	34	0.03
16.375	30	0.03	18.875	25	0.02
16.425	50	0.04	18.925	47	0.04
16.475	51	0.04	18.975	27	0.02
16.525	56	0.05	19.025	55	0.05
16.575	25	0.02	19.075	60	0.05
16.625	47	0.04	19.125	32	0.03
16.675	44	0.04	19.175	21	0.02
16.725	13	0.01	19.225	38	0.03
16.775	52	0.04	19.275	69	0.06
16.825	40	0.03	20.075	45	0.04
16.875	37	0.03	20.125	46	0.04
16.925	21	0.02	20.175	23	0.02
16.975	6	0.01	20.225	23	0.02
17.025	19	0.02	20.275	71	0.06
17.075	37	0.03	20.325	93	0.07
17.125	27	0.02	20.375	7	0.01

Middle of class h_i [Å]	Frequence f_i	Relative frequence f_i/n [%]	Middle of class h_i [Å]	Frequence f_i	Relative frequence f_i/n [%]
19.325	18	0.02	19.775	79	0.07
19.375	55	0.05	19.825	33	0.03
19.425	5	0.0	19.875	25	0.02
19.475	53	0.04	19.925	11	0.01
19.525	25	0.02	19.975	102	0.08
19.575	85	0.07	20.025	80	0.07
19.625	16	0.01	20.425	25	0.02
19.675	54	0.04	20.475	16	0.01
19.725	9	0.01	20.525	39	0.03

of structures of Ni^{2+} compounds are presented. Using program BIBSER, reference codes (REFCODE) of crystal structures were obtained. From this file, REFCODE of Ni^{2+} compounds were separated. Mixed — valency nickel compounds were excluded. This file of reference codes was further used for translation of DAT set by program RETRIEVE. In applying program GEOM, fractional coordinates of unique sets of nonhydrogen monatomic entities were elected by program UNIMOL. The computation of interatomic vector lengths with their origin in the position of central entity was performed by applying program DIRDOM, the algorithm of which is based on testing generated lengths of interatomic vectors with a common origin in the site of central entity. Vector lengths were generated up to the maximum value 30 Å. When the vector length of such a set is smaller or equal to the distance of its terminal point to any translation image if the position of central entity calculated by Eq. (1.2.1), then its terminal point is contained in the influence domain of the central entity position. The numbers u, v and w are modified in the case of a non-primitive unit cell. The length of such a vector is accepted for further computations. Calculated in this way the relative frequencies of interatomic vector lengths for monatomic entities of transition elements, are further graphically presented in the form of *histograms.*

The values of relative frequencies on the scales of these histograms are multiplied by constants listed in Table 1.2.2. The calculation of frequencies was performed only for those output sets, obtained by using programme GEOM, that contained coordinates of at least ten central atoms of a studied monatomic entity. The computations were made on computer EC 1033 at the Computing Centre of Chemical Technology Institute in Prague, using *Cambridge Crystallographic Database* [61].

Table 1.2.2. Multiplication factors (MF) by which the scales of frequencies in the histogramms of interatomic vector lengths are multiplied

Central monatomic entity	MF	Central monatomic entity	MF	Central monatomic entity	MF
Se^{3+}	100	Cu^{2+}	1	Sm^{3+}	1000
Ti^{2+}	100	Zn^{2+}	1	Eu^{3+}	1
Ti^{3+}	10	Y^{3+}	100	Gd^{3+}	1
Ti^{4+}	1	Zr^{2+}	100	Er^{3+}	1
V	10	Zr^{4+}	1	Yb^{3+}	100
V^{1+}	100	Nb^{1+}	100	Lu^{3+}	1000
V^{2+}	100	Nb^{2+}	100	Hf^{4+}	100
V^{3+}	10	Nb^{3+}	1000	Ta^{3+}	1
V^{4+}	100	Nb^{5+}	10	Ta^{5+}	1
V^{5+}	100	Mo	100	W	10
Cr	10	Mo^{1+}	100	W^{1+}	1000
Cr^{1+}	100	Mo^{2+}	100	W^{2+}	100
Cr^{2+}	10	Mo^{3+}	100	W^{3+}	100
Cr^{3+}	1	Mo^{4+}	10	W^{4+}	100
Mn^{1-}	1	Mo^{5+}	1	W^{5+}	100
Mn	1	Mo^{6+}	1	W^{6+}	10
Mn^{1+}	1	Tc^{3+}	1000	Re	1000
Mn^{2+}	1	Tc^{5+}	100	Re^{1+}	10
Mn^{3+}	100	Ru	10	Re^{2+}	100
Fe^{2-}	100	Ru^{1+}	100	Re^{3+}	10
Fe	1	Ru^{2+}	1	Re^{4+}	100
Fe^{1+}	1	Ru^{3+}	1000	Re^{5+}	100
Fe^{2+}	1	Rh	1	Os	1
Fe^{3+}	1	Rh^{1+}	1	Os^{2+}	10
Co^{1-}	1	Rh^{2+}	100	Os^{6+}	10
Co	10	Rh^{3+}	100	Ir	10
Co^{1+}	1	Pd	100	Ir^{1+}	100
Co^{2+}	1	Pd^{1+}	1	Ir^{2+}	100
Co^{3+}	1	Pd^{2+}	10	Ir^{3+}	1
Ni	10	Ag^{1+}	1	Pt	1
Ni^{1+}	1	Cd^{2+}	1	Pt^{2+}	1
Ni^{2+}	1	Lu^{3+}	1	Pt^{4+}	10
Ni^{3+}	1	Ce^{3+}	1	Au^{1+}	1
Ni^{4+}	1	Ce^{4+}	1	Au^{3+}	100
Cu^{1+}	1	Pr^{3+}	10	Hg^{1+}	100
		Nd^{3+}	1	Hg^{2+}	1

RELATIVE FREQUENCE [%]

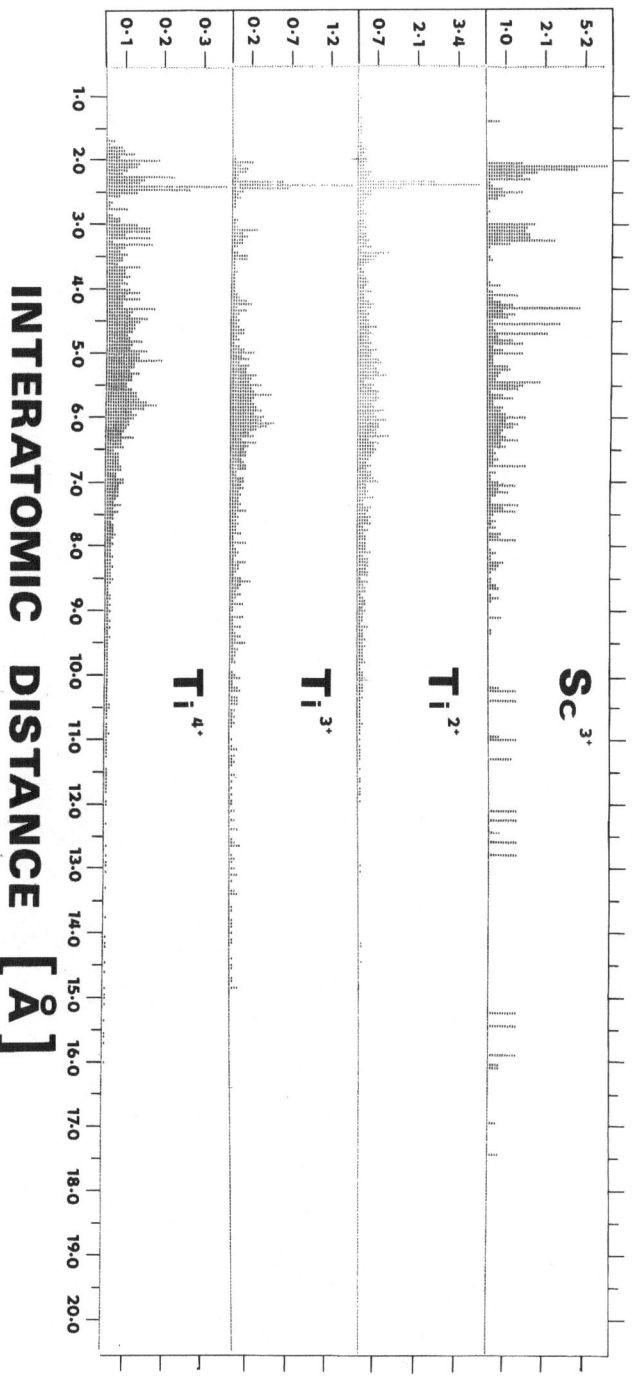

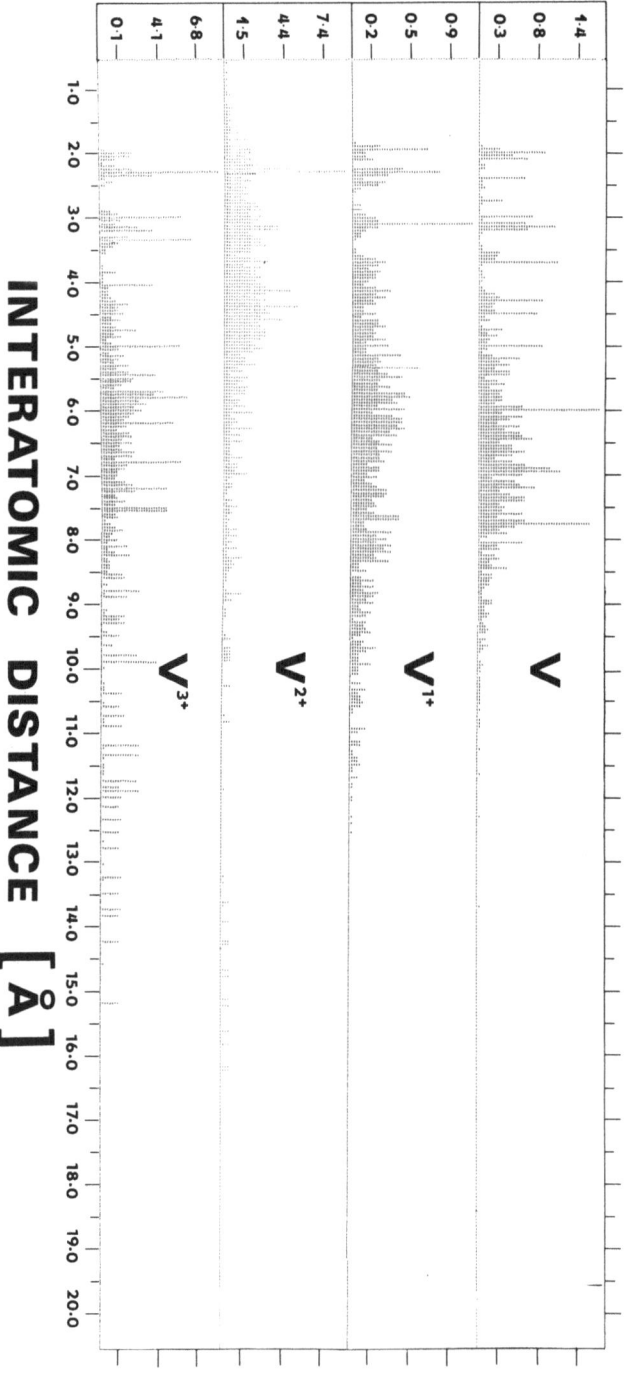

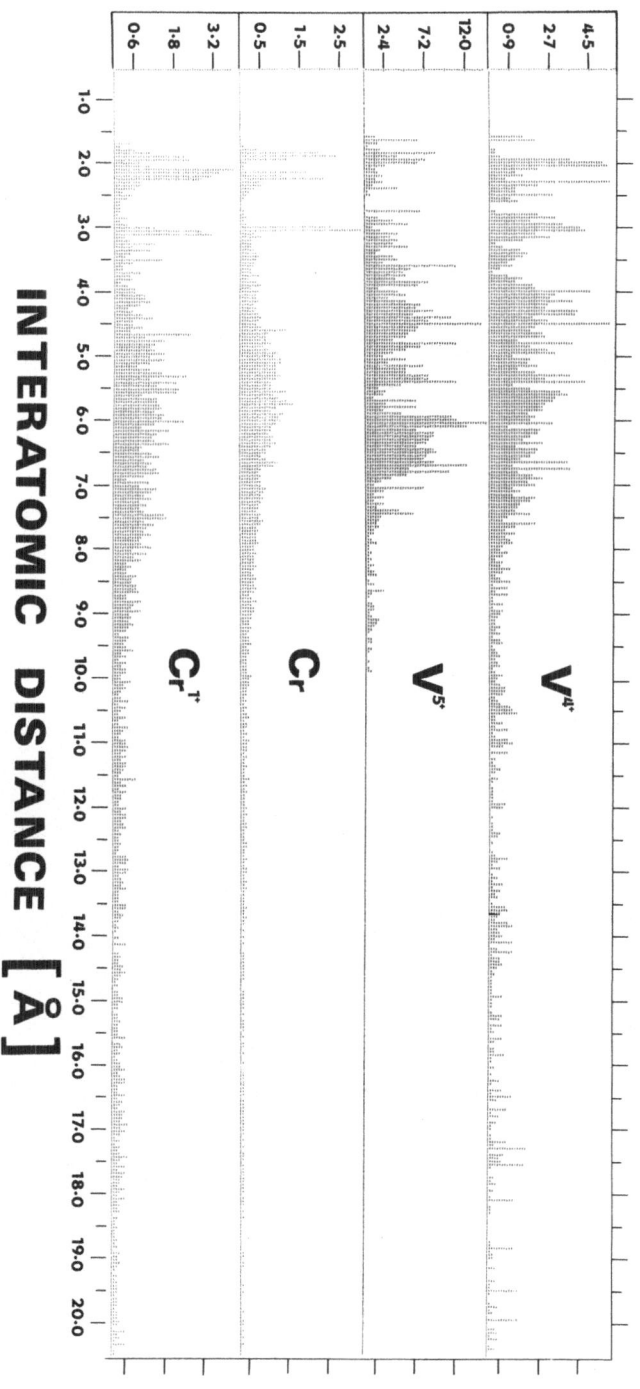

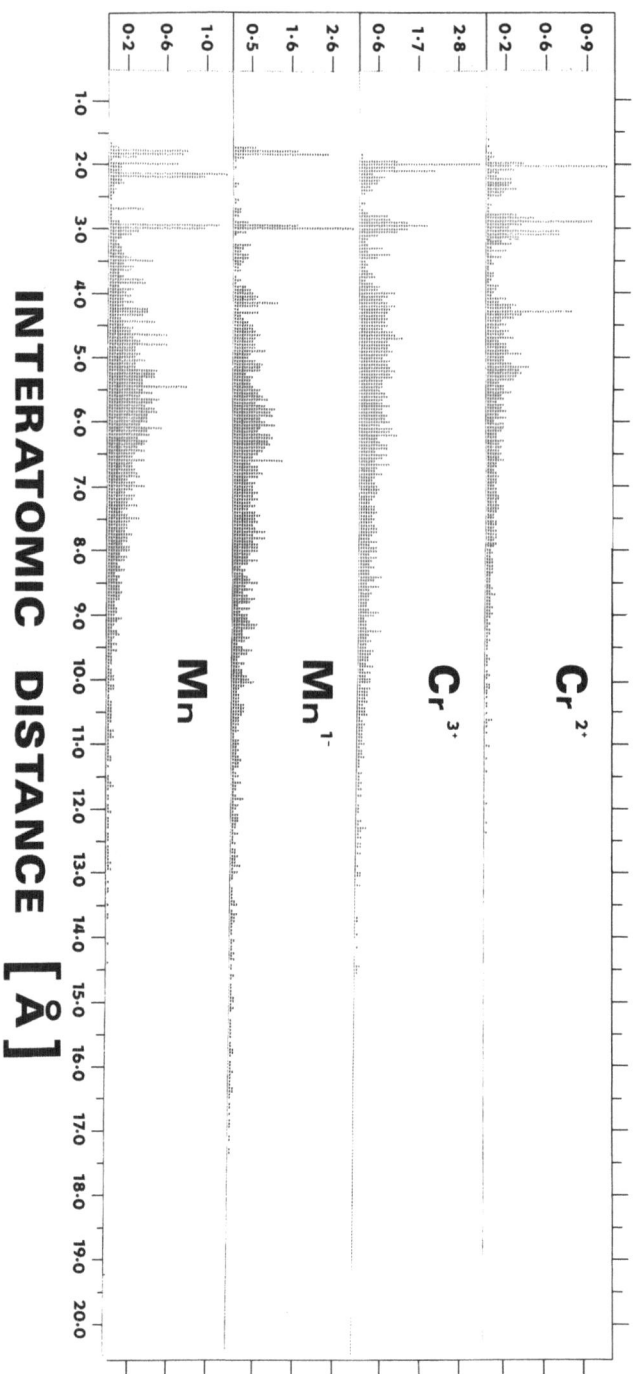

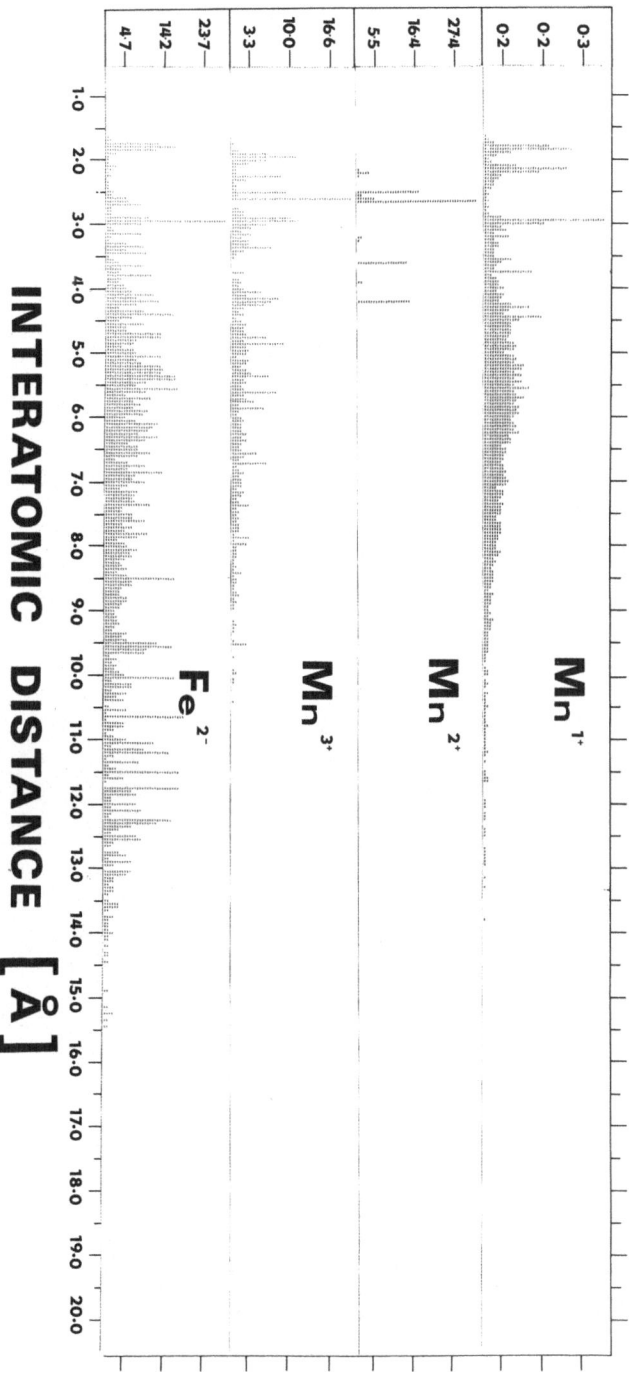

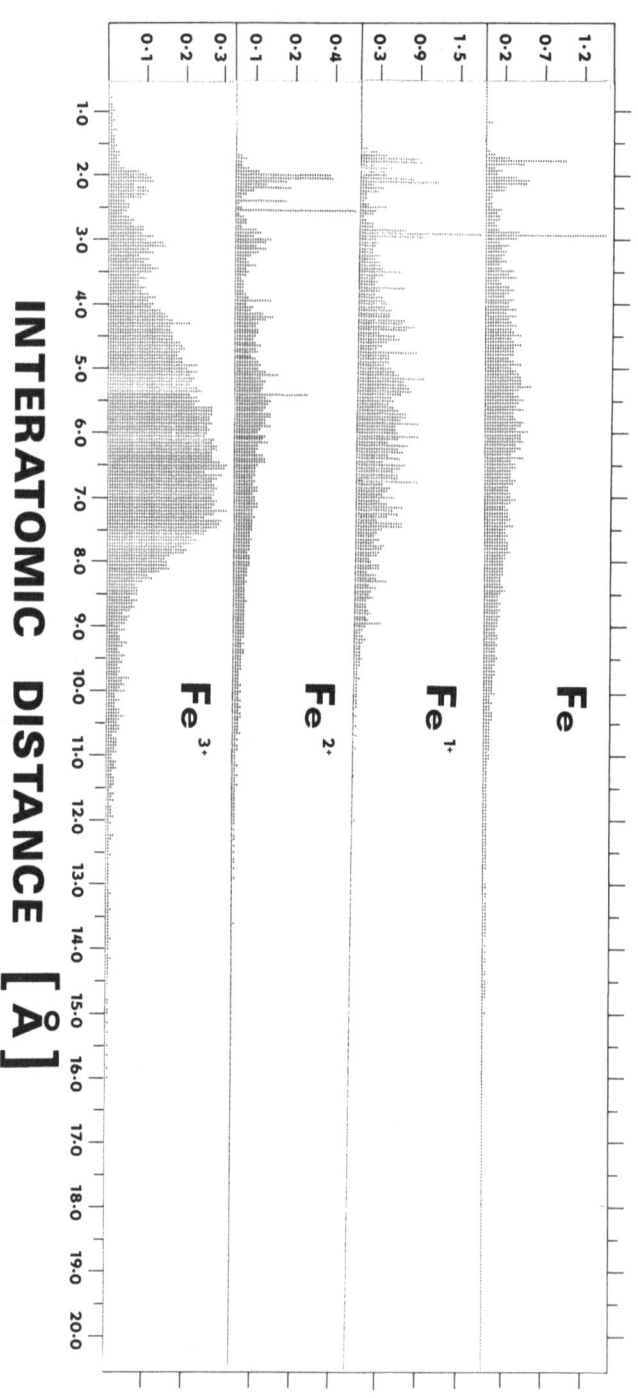

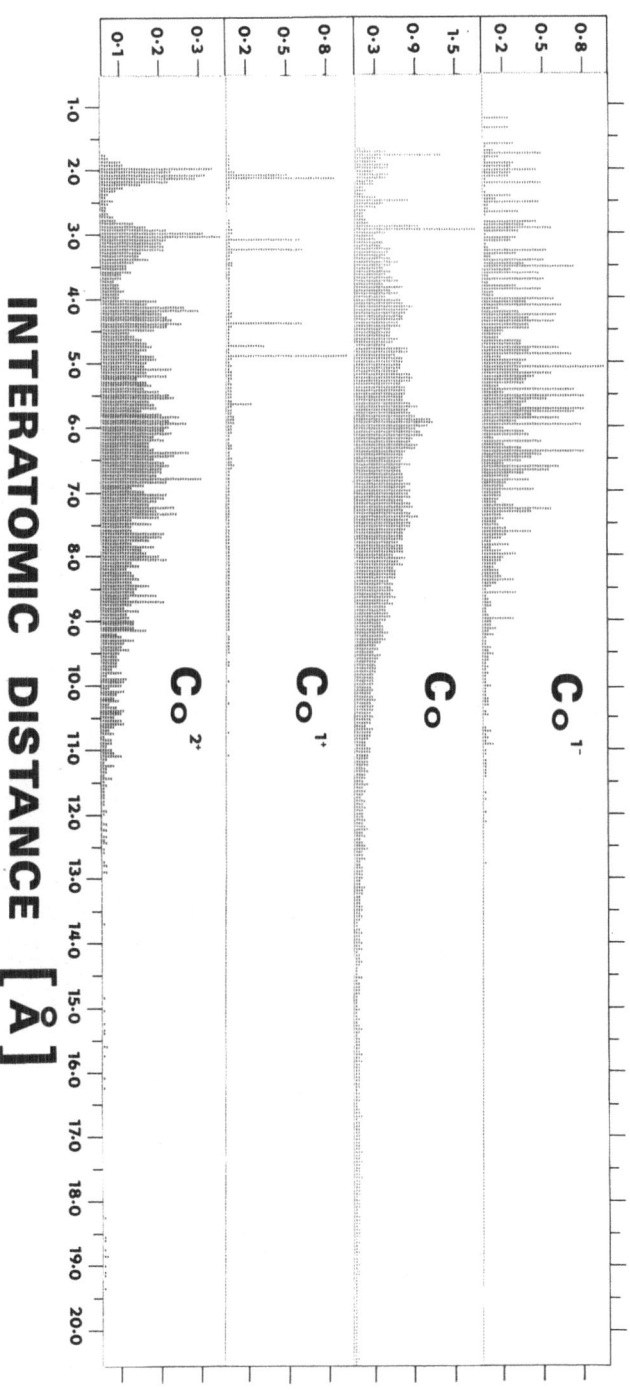

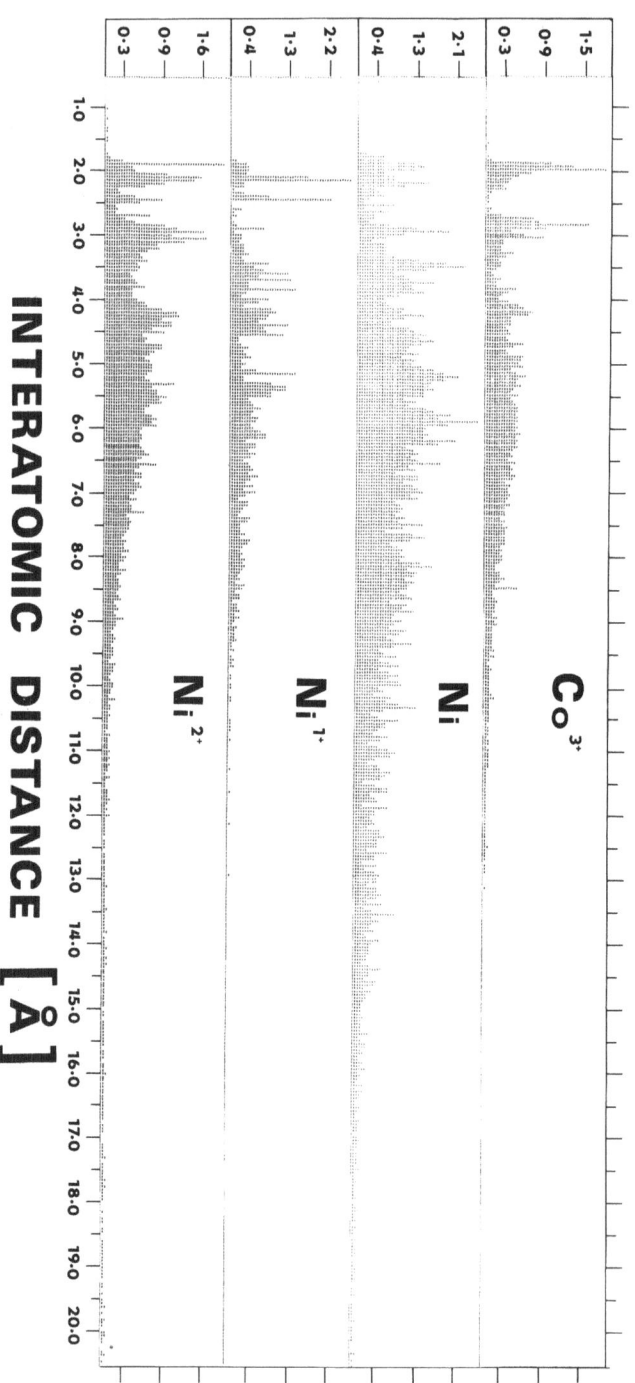

RELATIVE FREQUENCE [%]

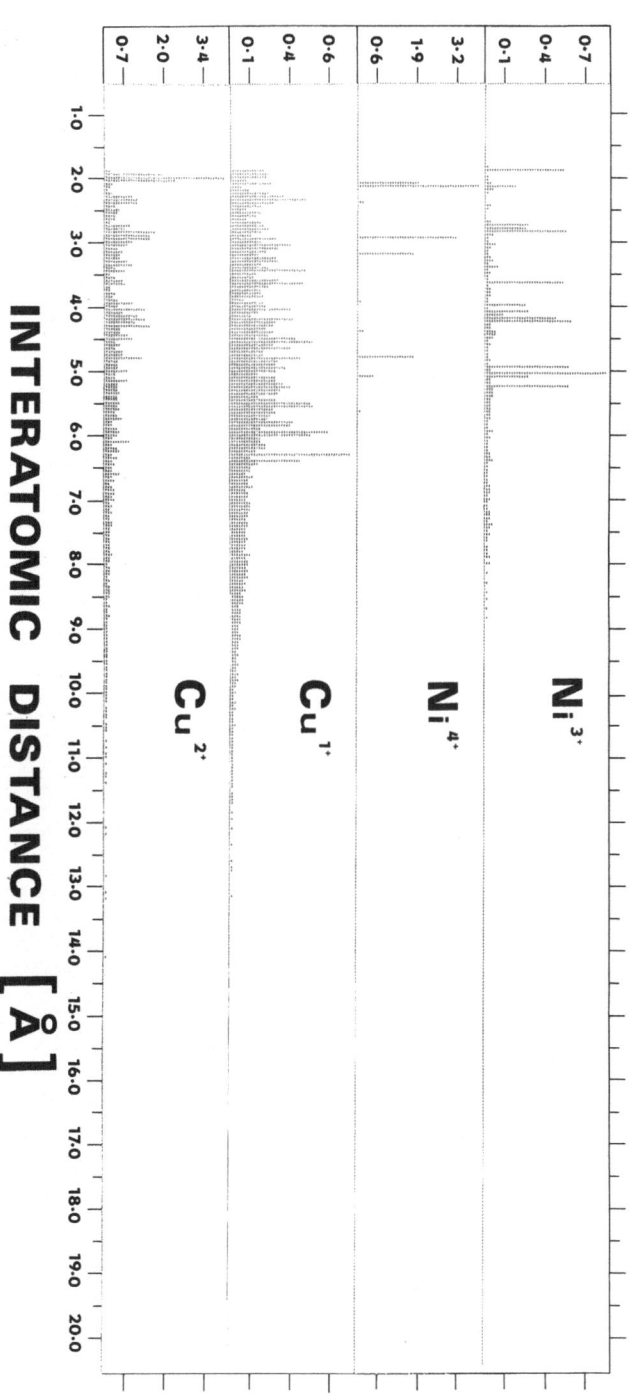

INTERATOMIC DISTANCE [Å]

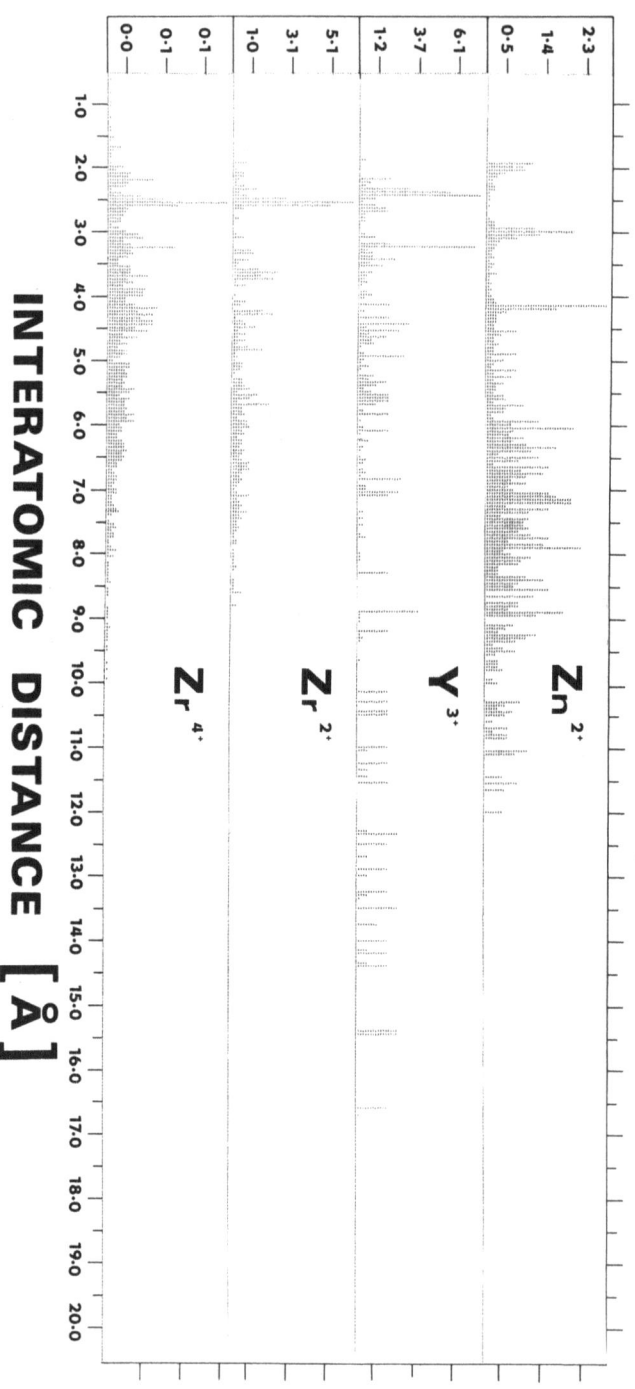

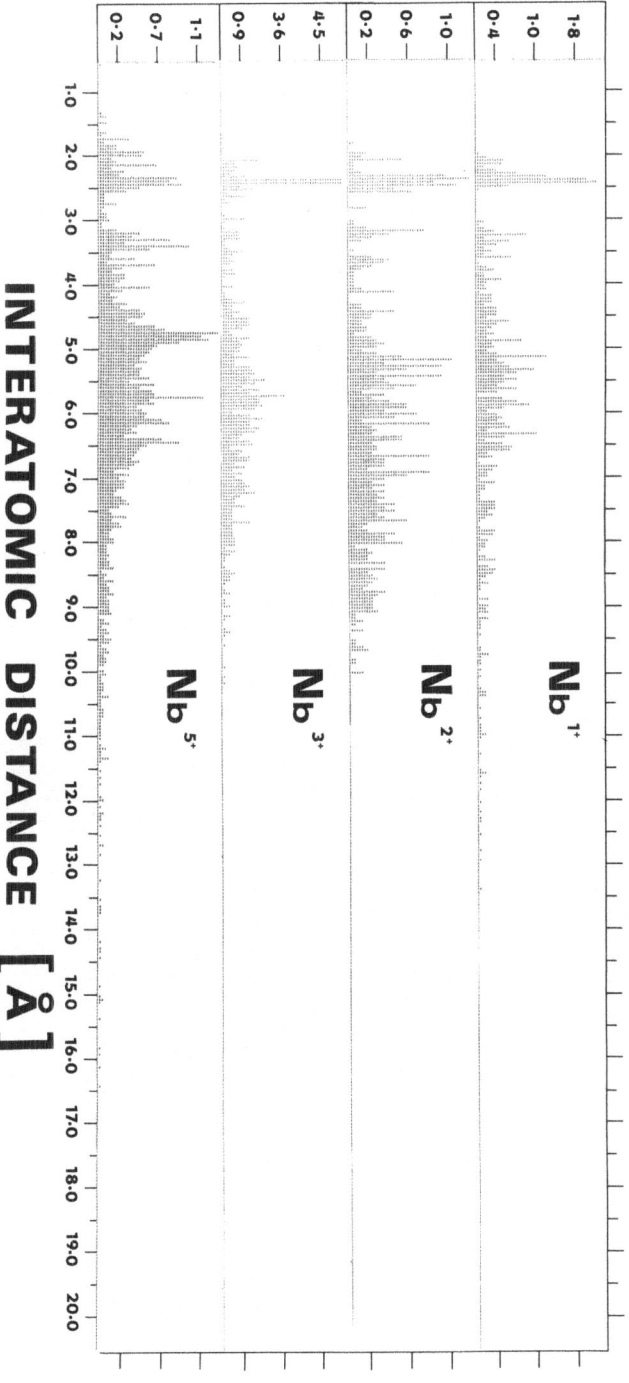

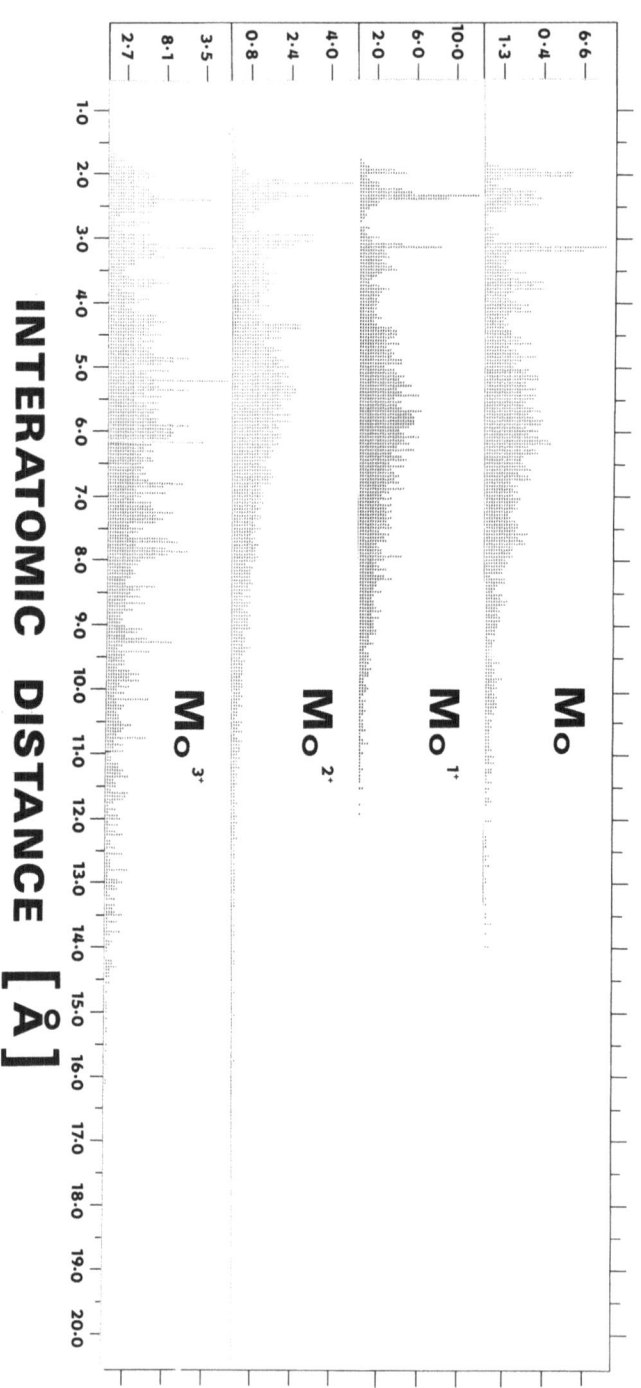

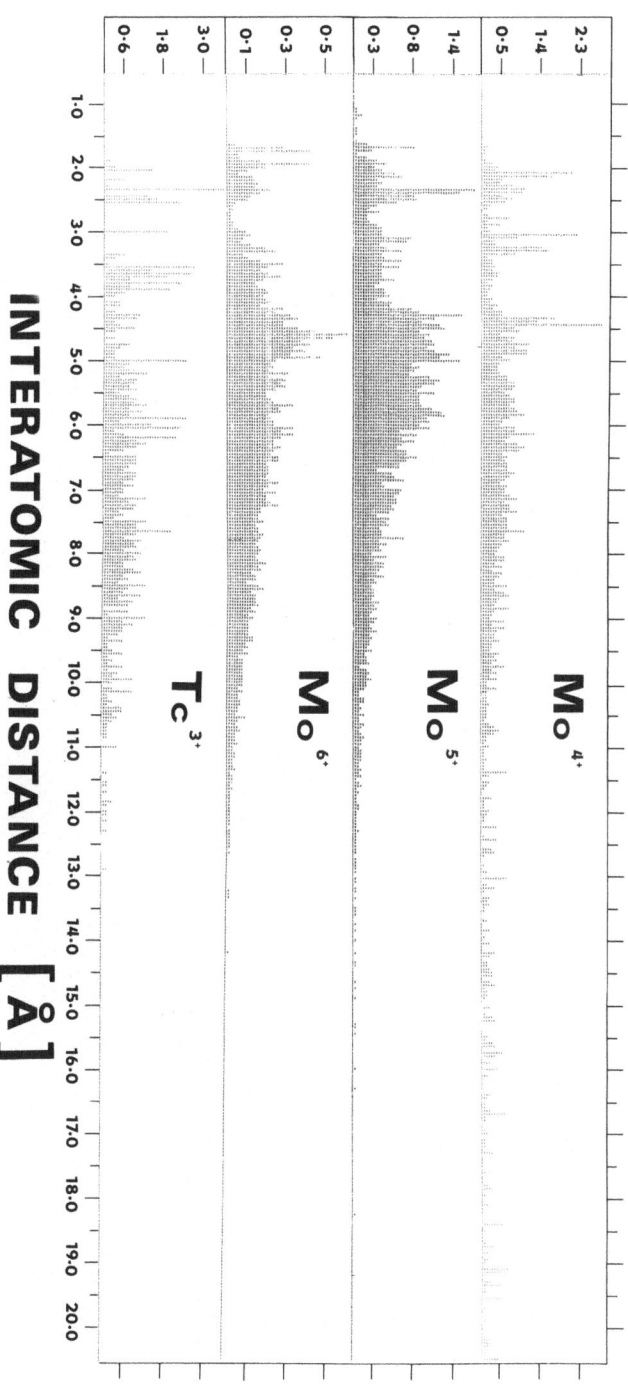

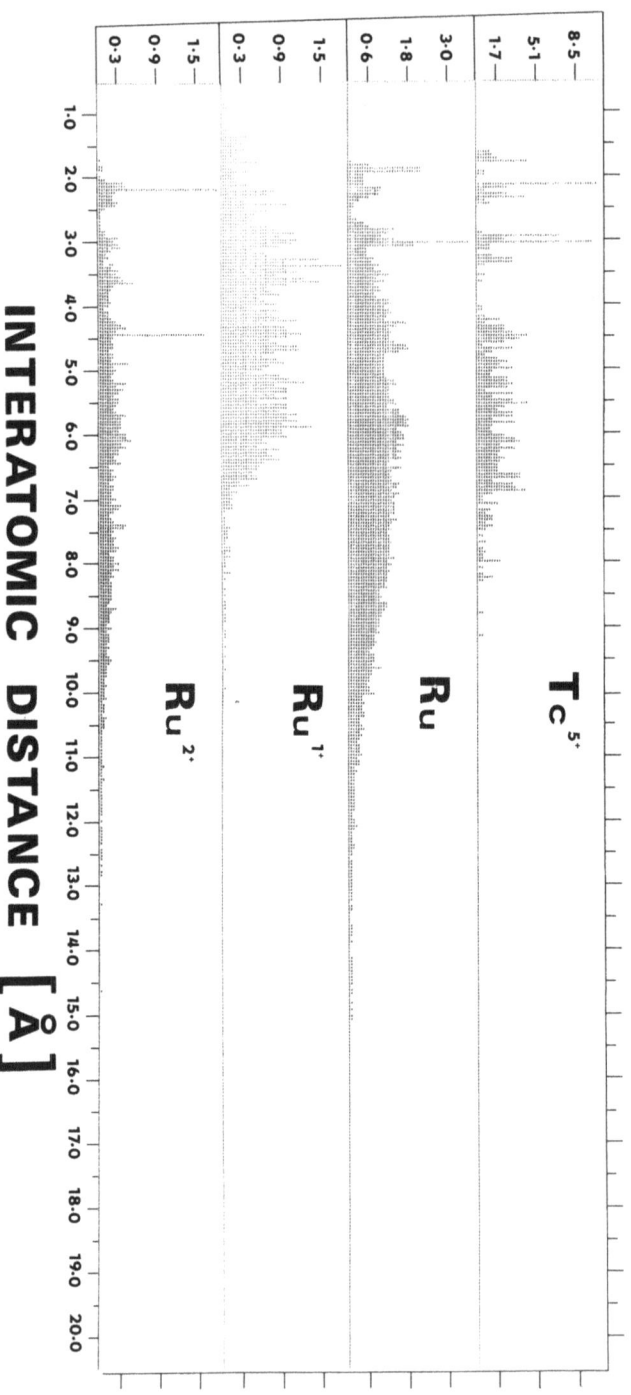

RELATIVE FREQUENCE [%]

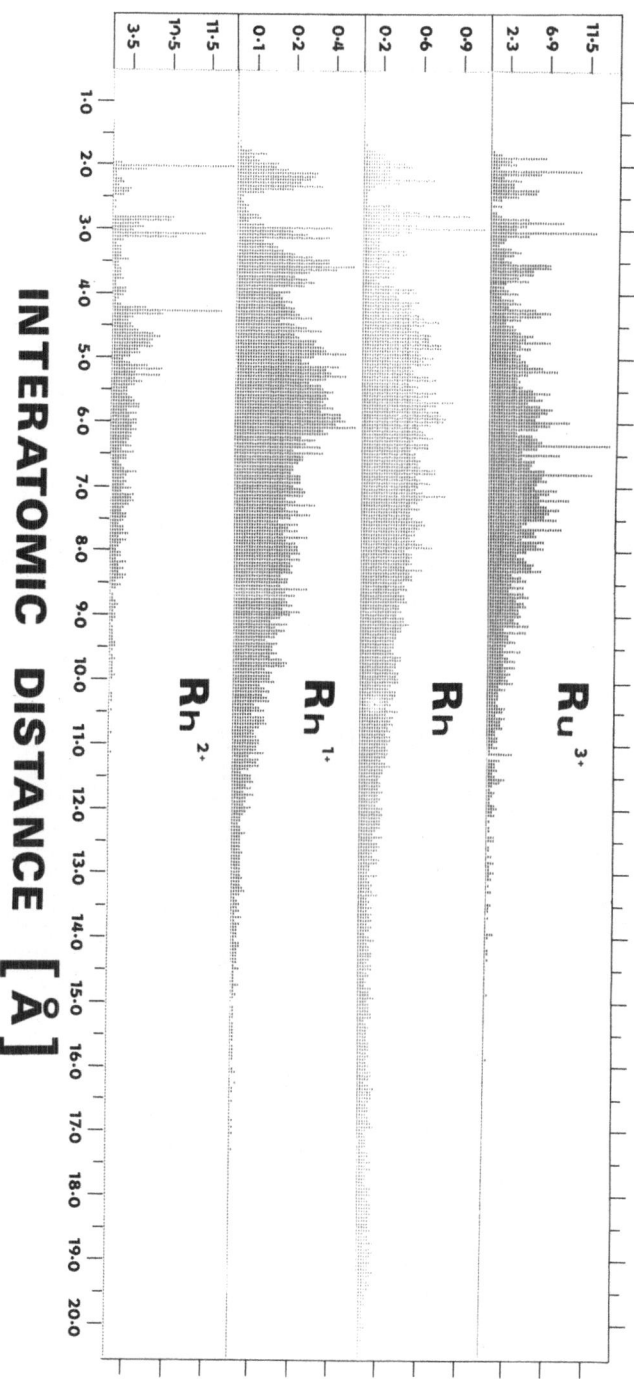

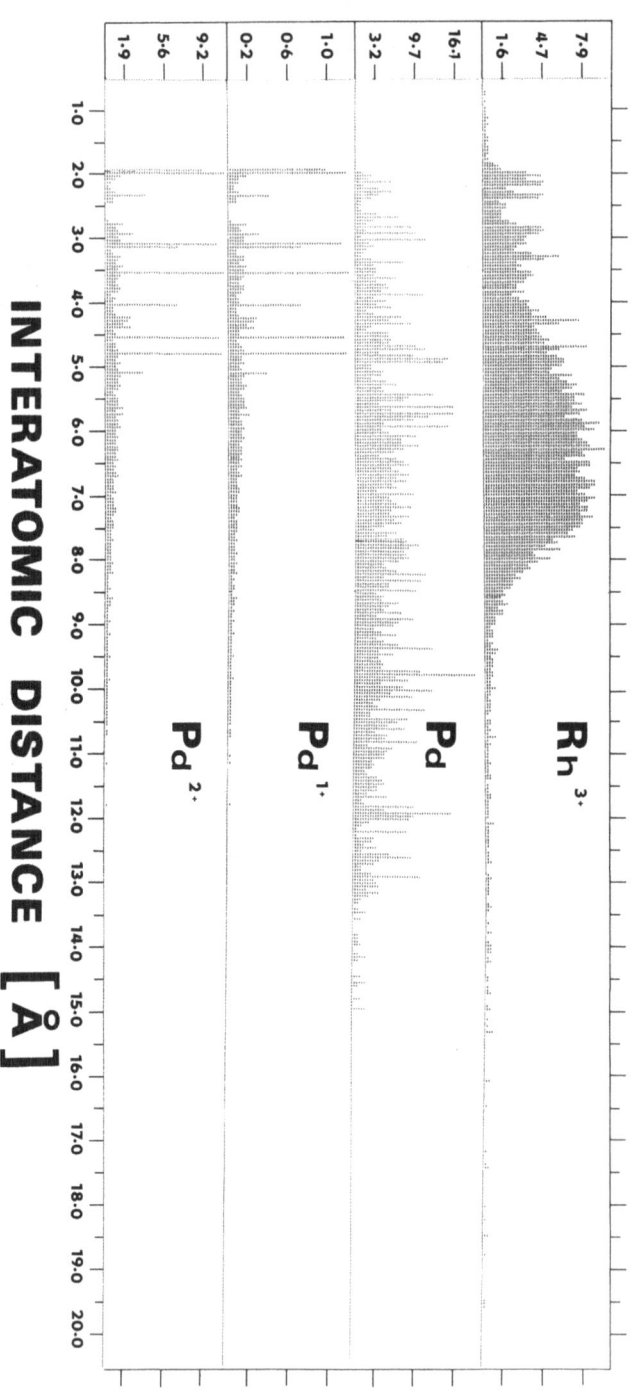

RELATIVE FREQUENCE [%]

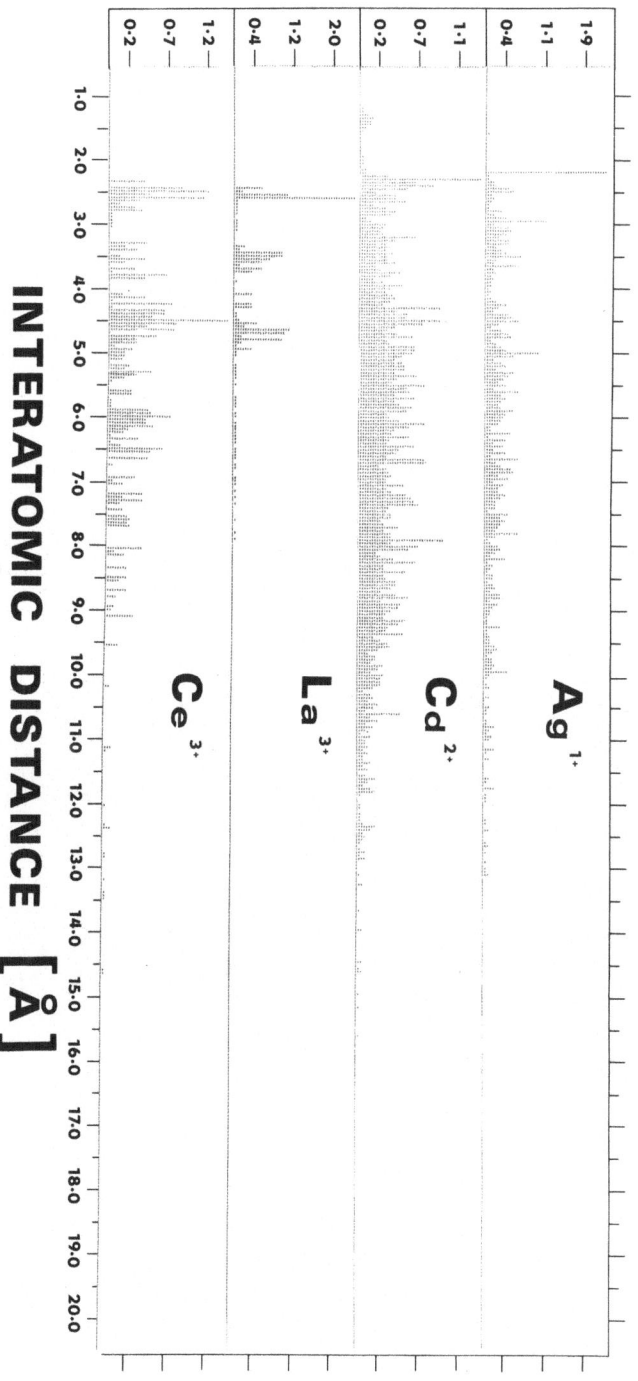

INTERATOMIC DISTANCE [Å]

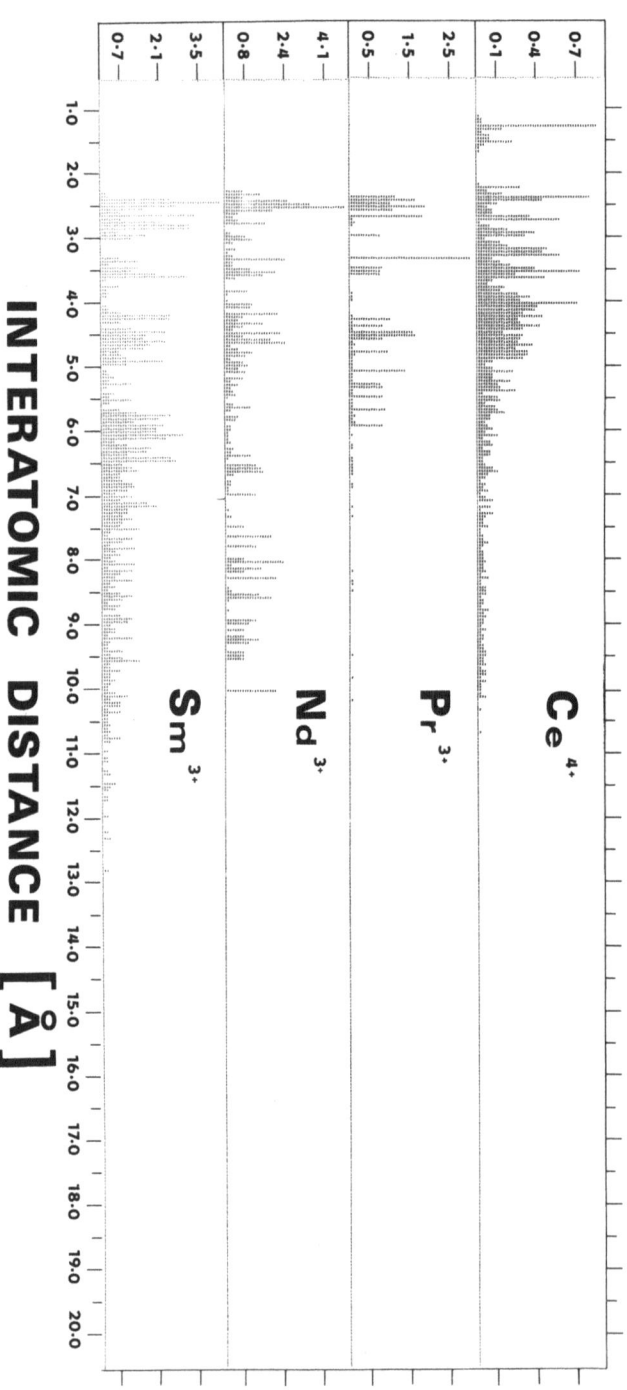

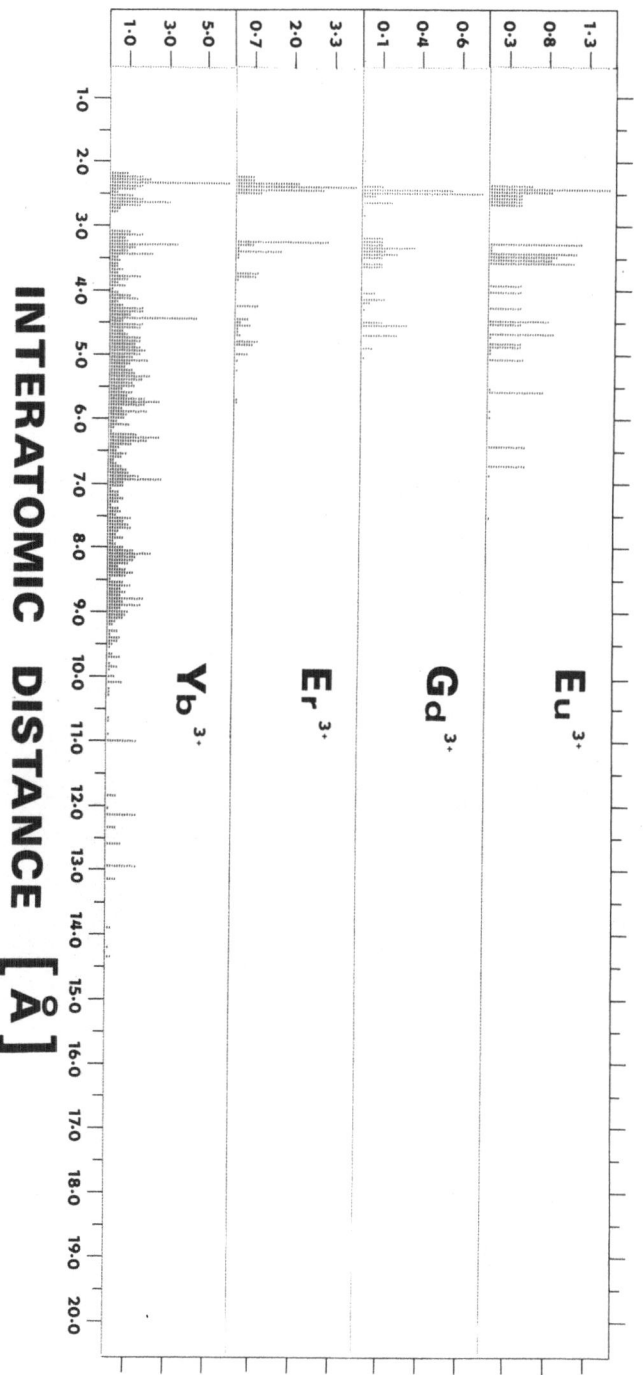

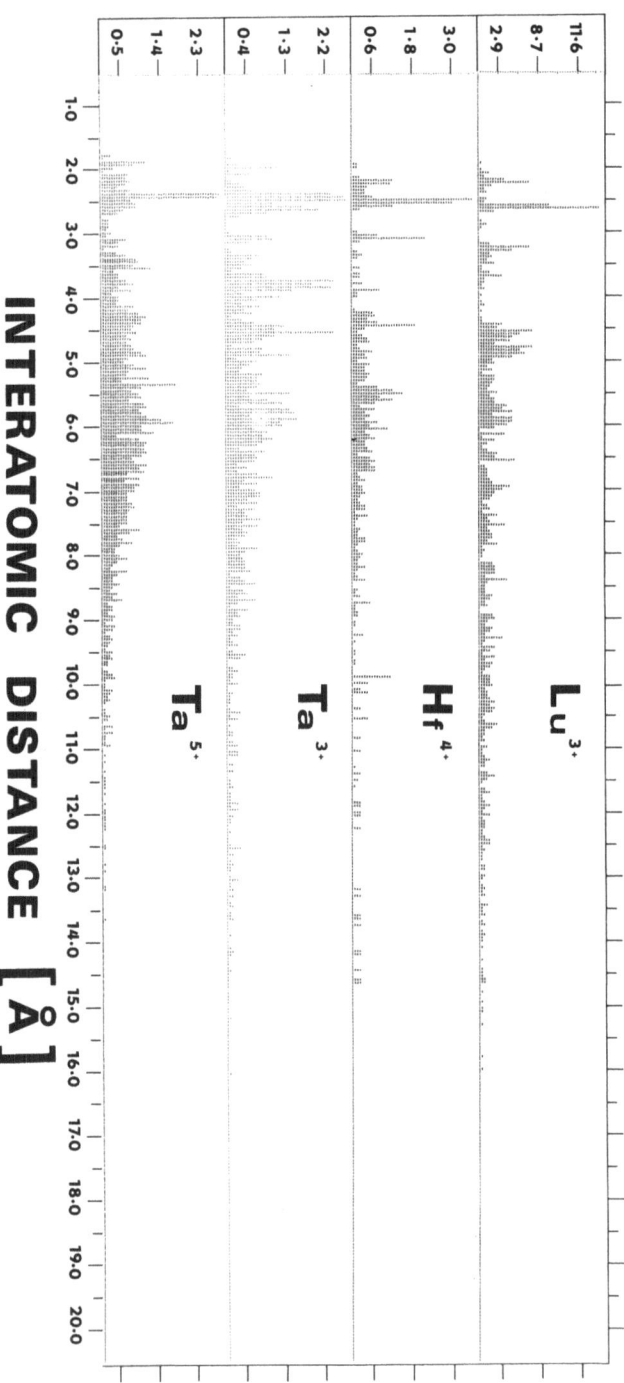

RELATIVE FREQUENCE [%]

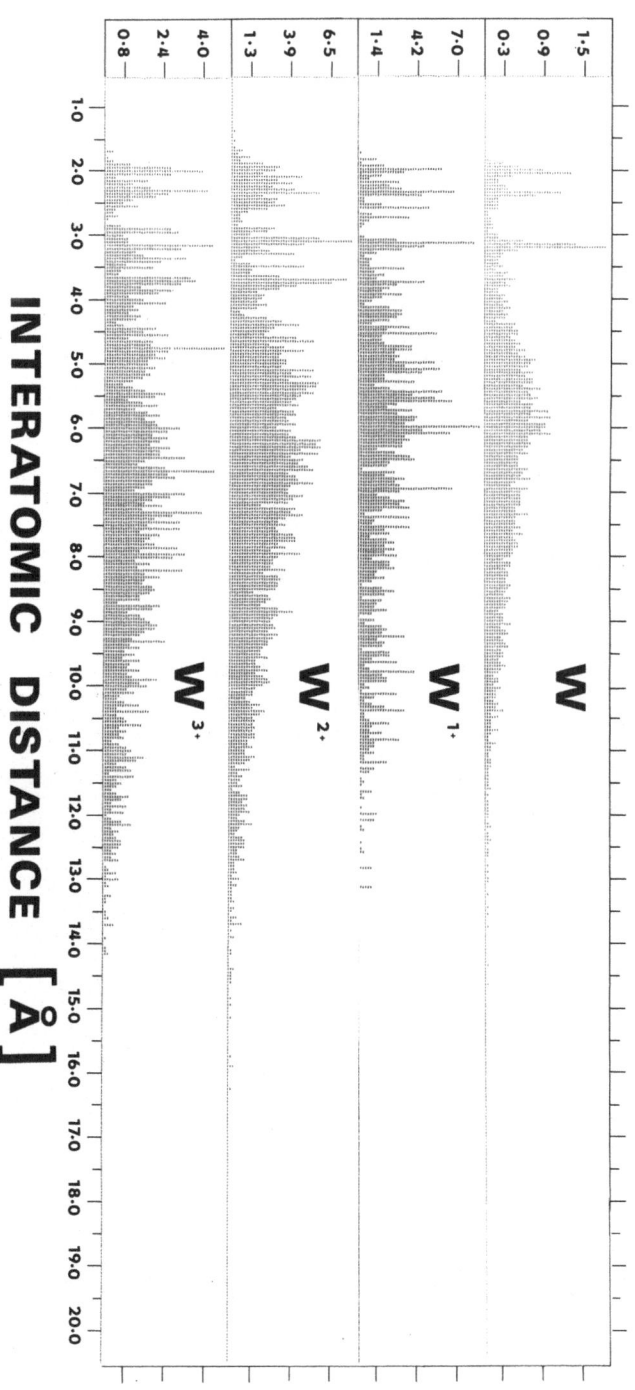

RELATIVE FREQUENCE [%]

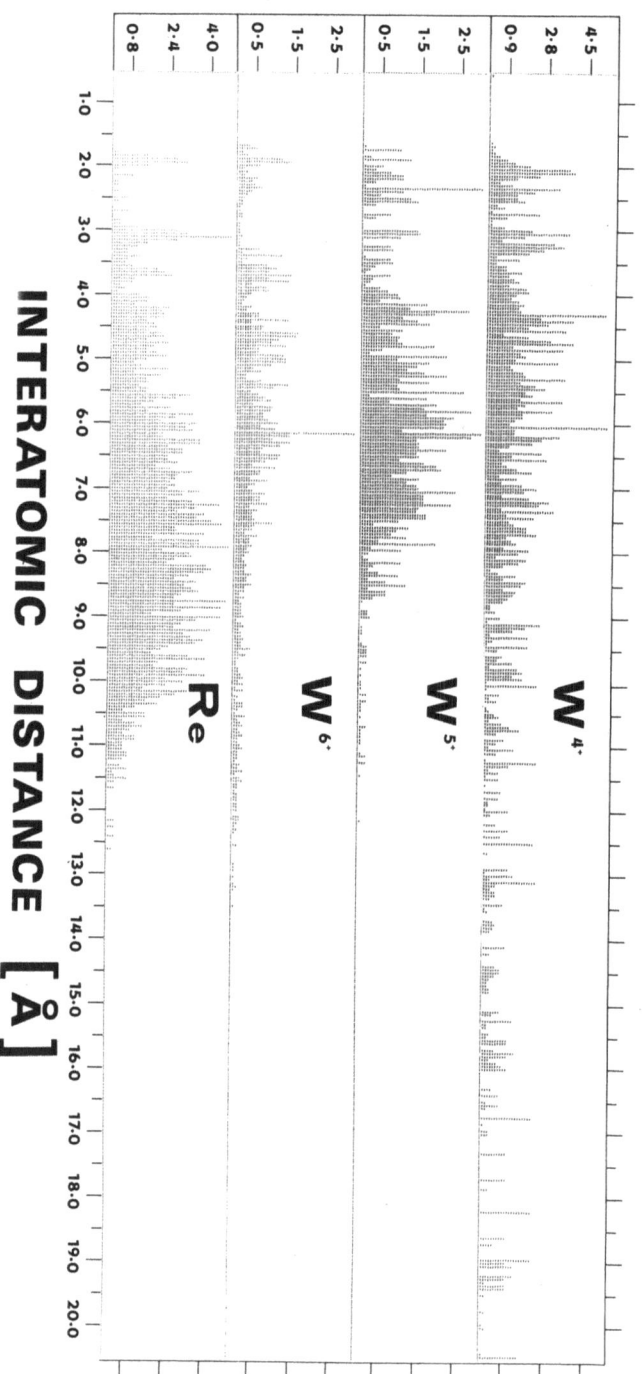

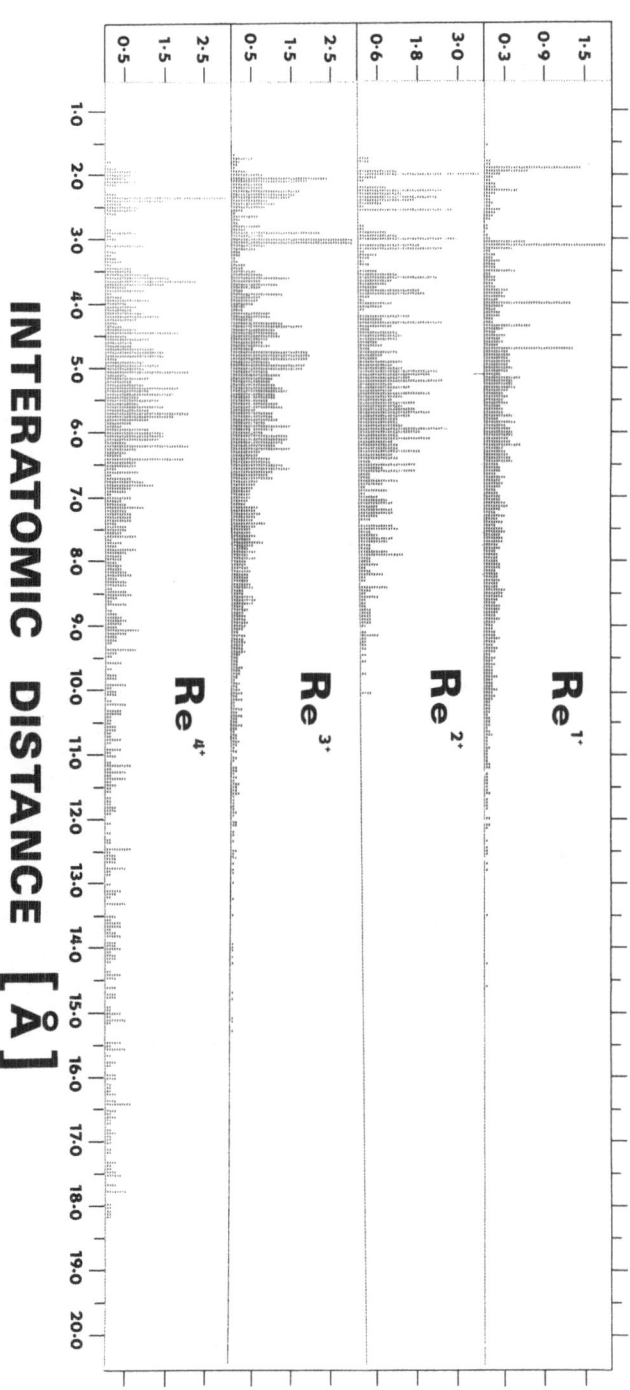

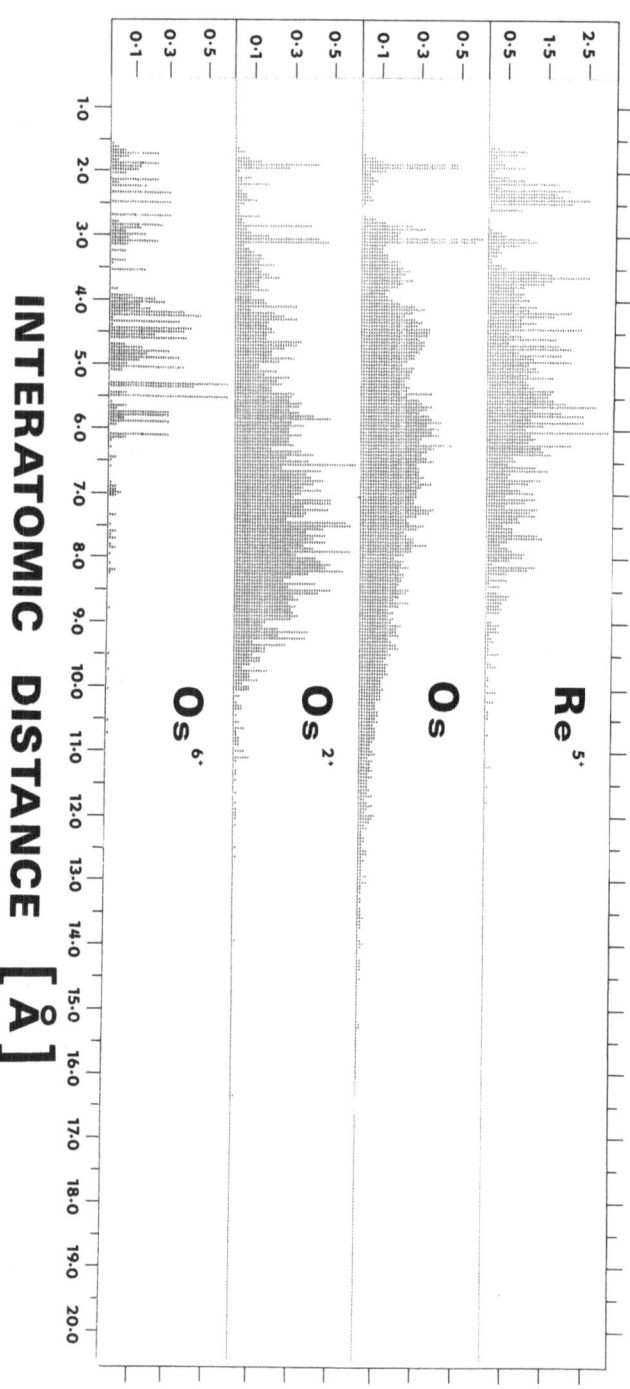

RELATIVE FREQUENCE [%]

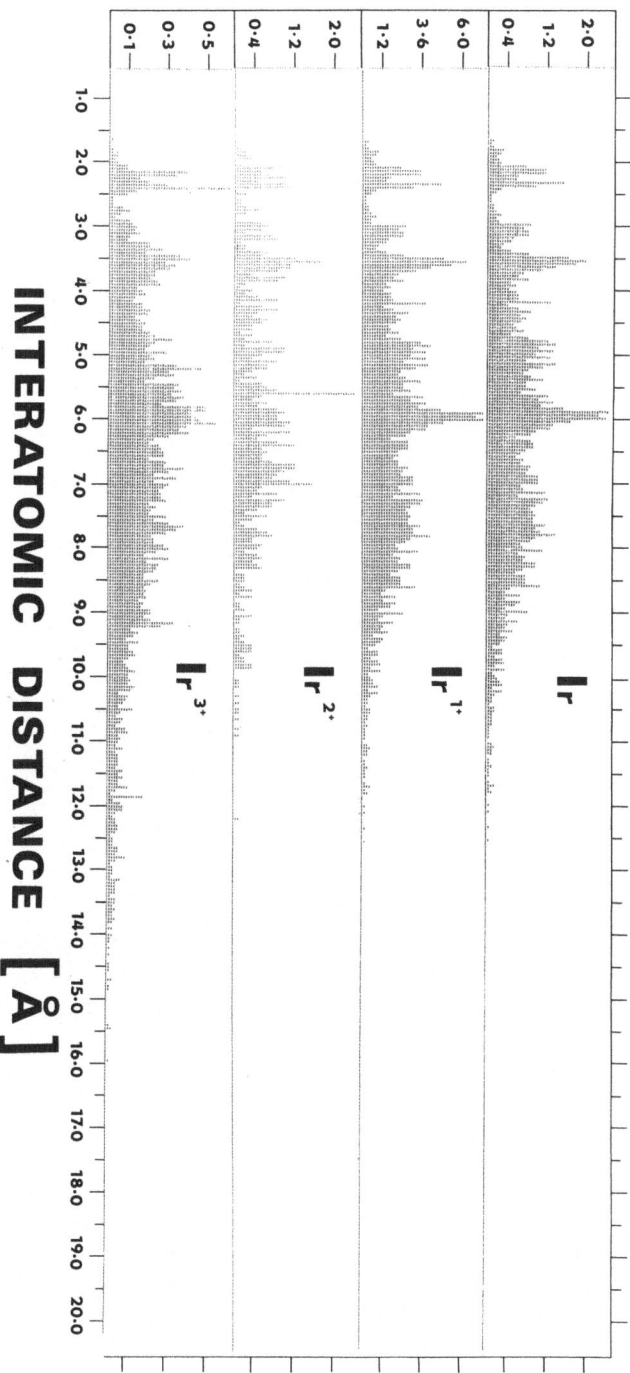

INTERATOMIC DISTANCE [Å]

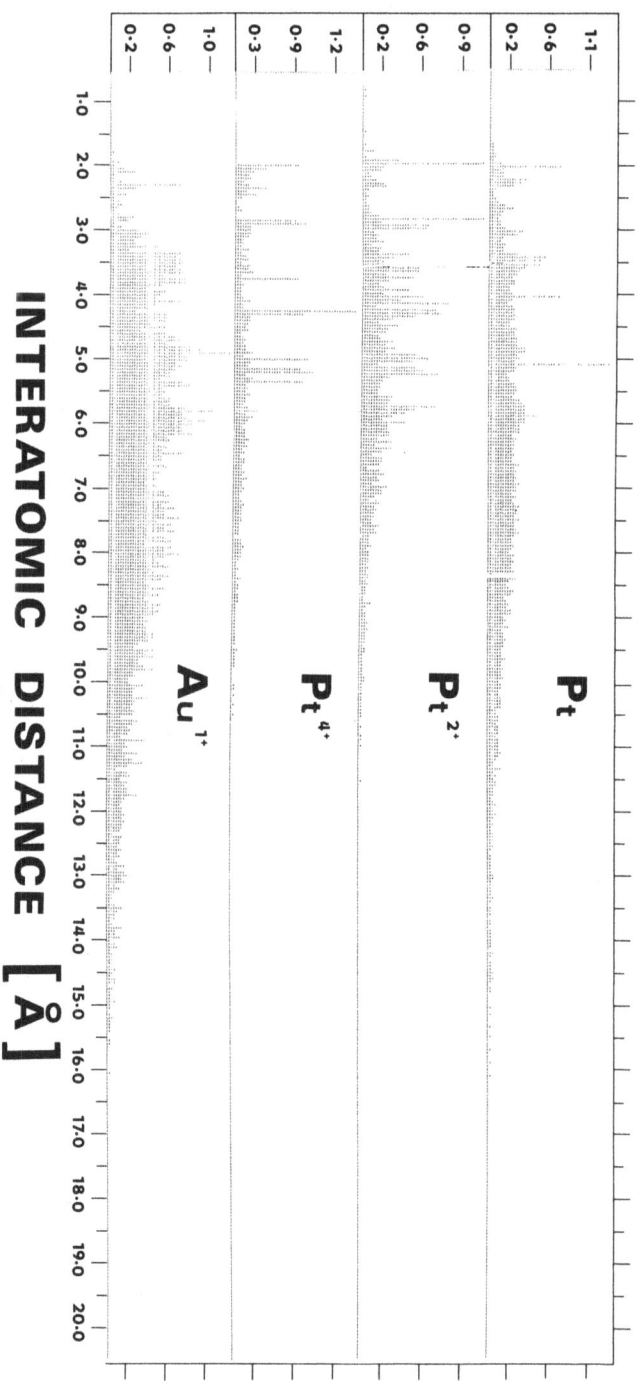

RELATIVE FREQUENCE [%]

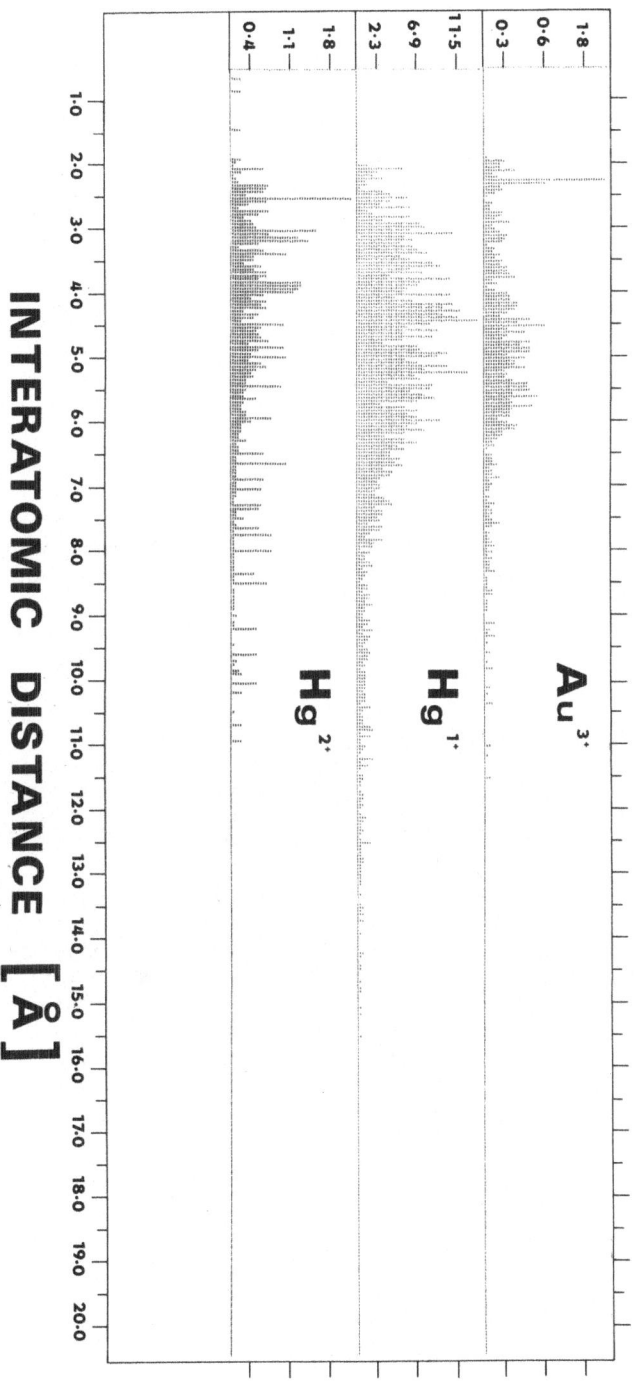

INTERATOMIC DISTANCE [Å]

In Table 1.2.3 are collected empirical estimates of the first two central moments and asymmetries (1.1.4) of these empirical distributions. To calculate the estimates of the first three central moments the formulas below were used:

$$m_1 = \sum_{i=1}^{l} \left[f_i(g_u + i\Delta) - \frac{\Delta}{2} \right] \bigg/ n$$

$$m_2 = \sum_{i=1}^{l} \left[f_i(g_u + i\Delta) - \frac{\Delta}{2} - m_1 \right]^2 \bigg/ n \qquad (1.2.2)$$

$$m_3 = \sum_{i=1}^{l} \left[f_i(g_u + i\Delta) - \frac{\Delta}{2} - m_1 \right]^3 \bigg/ n$$

where g_u is the lower limit of the class containing the value of x_1. The asymmetry estimates were computed using formula

$$\gamma = \frac{m_3}{m_2^{3/2}}$$

These computations were made by applying programs of bank MEB [62] with computer MINSK 4030.1 at the Computing Centre of Slovak Technical University in Bratislava.

Table 1.2.3. Empirical estimates of the first two central moments m_1, m_2 and asymmetries γ of the interatomic vector lengths distributions. The sample sizes are n; n_{crit} is the critical value of sample size calculated by formula (1.1.8). The significance level is $\alpha = 0.05$

Central monatomic entity	m_1 [Å]	m_2 [Å²]	γ	n	n_{crit}
Sc^{3+}	5.801	11.883	1.330	254 566	18 260
Ti^{2+}	5.720	6.937	0.583	665 293	10 660
Ti^{3+}	6.353	9.630	0.941	30 769	14 798
Ti^{4+}	5.070	6.847	1.512	163 162	10 521
V	5.913	3.773	−0.302	54 211	5 798
V^{1+}	6.065	5.405	0.133	347 078	8 306
V^{2+}	4.617	6.280	2.068	82 419	9 650
V^{3+}	6.516	8.612	0.649	9 980	13 234
V^{4+}	6.723	18.954	1.390	146 480	29 125
V^{5+}	5.065	2.847	−0.068	147 996	4 375
Cr	6.417	13.491	1.409	131 273	20 731
Cr^{1+}	7.510	20.734	1.030	247 281	31 861
Cr^{2+}	4.897	4.804	0.754	45 042	7 382
Cr^{3+}	5.860	7.318	0.612	15 447	11 245
Mn^{1-}	7.048	10.813	0.778	4 142	16 616
Mn	5.524	5.992	0.673	28 795	9 208
Mn^{1+}	5.343	5.413	0.614	149 921	8 318
Mn^{2+}	3.169	0.951	2.498	192 798	1 461
Mn^{3+}	4.664	3.948	0.499	92 655	6 067
Fe^{2-}	7.378	10.423	0.236	96 320	16 016

Central monatomic entity	m_1 [Å]	m_2 [Å²]	γ	n	n_{crit}
Fe	5.980	7.992	1.105	107840	12281
Fe^{1+}	5.411	4.272	0.144	32276	6565
Fe^{2+}	5.659	7.311	0.892	461373	11234
Fe^{3+}	6.136	4.998	0.751	452367	7680
Co^{1-}	5.313	3.997	0.339	25600	6142
Co	7.188	13.871	1.225	288012	21315
Co^{1+}	4.905	7.353	1.849	319700	11299
Co^{2+}	6.077	6.643	0.791	400497	10208
Co^{3+}	5.565	6.786	0.778	106212	10428
Ni	7.657	14.440	0.786	368468	22189
Ni^{1+}	5.159	4.368	0.536	7423	6712
Ni^{2+}	5.853	12.087	1.773	120161	18573
Ni^{3+}	4.376	2.263	0.503	51715	3477
Ni^{4+}	3.063	1.389	1.622	139559	2134
Cu^{1+}	5.176	4.328	0.847	193553	6651
Cu^{2+}	4.863	7.717	1.847	449545	11858
Zn^{2+}	6.873	5.269	−0.200	118219	8097
Y^{3+}	7.073	17.617	0.646	196426	27071
Zr^{2+}	4.634	3.166	0.286	197900	4865
Zr^{4+}	4.437	3.225	0.929	253150	4956
Nb^{1+}	5.363	5.123	0.664	161570	7872
Nb^{2+}	5.729	4.132	−0.205	162333	6349
Nb^{3+}	5.305	3.596	−0.097	659873	5526
Nb^{5+}	5.396	5.078	0.970	167505	7803
Mo	5.749	5,803	0.553	241977	8917
Mo^{1+}	5.751	4.933	0.316	180479	7580
Mo^{2+}	5.621	6.229	0.867	1331649	9572
Mo^{3+}	6.454	8.672	0.613	103500	13326
Mo^{4+}	7.396	18.994	1.198	185296	29187
Mo^{5+}	5.685	6.029	0.977	15633	9264
Mo^{6+}	5.886	5.612	0.469	85907	8624
Tc^{3+}	6.204	5.467	0.121	660735	8401
Tc^{5+}	4.854	3.042	−0.217	31394	4674
Ru	6.322	6.393	0.402	185508	9824
Ru^{1+}	4.433	2.257	0.153	702908	3468
Ru^{2+}	5.979	6.657	0.592	45416	10229
Ru^{3+}	6.397	6.134	0.249	362249	9426
Rh	8.109	17.605	0.957	350132	27051
Rh^{1+}	-6.590	7.946	0.665	47049	12210
Rh^{2+}	5.147	3.754	0.386	120040	5769
Rh^{3+}	5.935	5.370	1.360	345800	8252
Pd	7.727	8.883	1.216	128304	13650
Pd^{1+}	4.807	5.009	1.260	204456	7697
Pd^{2+}	4.763	5.011	1.325	141767	7700
Ag^{1+}	6.295	10.634	1.210	1302574	16341
Cd^{2+}	6.405	7.020	0.417	656793	10787
La^{3+}	4.252	3.395	2.186	145918	5217
Ce^{3+}	5.202	4.308	1.022	88300	6620
Ce^{4+}	4.316	3.567	0.960	359476	5481
Pr^{3+}	4.227	2.938	1.601	268811	4515
Nd^{3+}	5.351	5.766	0.470	100419	8860
Sm^{3+}	5.784	5.618	0.384	666324	8633

Table 1.2.3 (continued)

Central monatomic entity	m_1 [Å]	m_2 [Å²]	γ	n	n_{crit}
Eu^{3+}	3.947	2.096	1.598	120212	3221
Gd^{3+}	3.399	1.479	3.445	694151	2273
Er^{3+}	3.326	1.655	2.123	251137	2543
Yb^{3+}	5.859	6.464	0.660	259507	9933
Lu^{3+}	6.728	10.102	0.548	255551	15523
Hf^{4+}	5.696	8.128	1.041	193919	12490
Ta^{3+}	5.506	6.399	0.987	289568	9833
Ta^{5+}	5.520	4.773	9.384	2330	7334
W	6.038	5.627	0.321	111926	8647
W^{1+}	6.138	6.457	0.356	463296	9922
W^{2+}	5.506	7.342	0.364	114547	11282
W^{3+}	6.594	7.761	0.240	153159	11926
W^{4+}	7.713	20.606	0.791	309117	31664
W^{5+}	5.650	3.373	−0.085	121880	5183
W^{6+}	5.996	5.746	0.585	117154	8830
Re	7.149	5.146	−0.387	696766	7908
Re^{1+}	5.849	6.123	0.435	152201	9409
Re^{2+}	5.077	3.466	0.119	59005	5326
Re^{3+}	5.402	5.560	0.729	60416	8544
Re^{4+}	7.064	14.879	1.121	89363	22864
Re^{5+}	5.267	3.475	0.109	164625	5340
Os	6.421	6.248	0.438	136321	9601
Os^{2+}	6.517	4.073	−0.367	158181	6259
Os^{6+}	4.335	2.218	0.306	52895	3408
Ir	5.989	4.616	0.148	52184	7093
Ir^{1+}	5.979	4.572	0.149	169904	7026
Ir^{2+}	5.605	4.109	0.141	93409	6314
Ir^{3+}	6.599	7.050	0.505	22826	10833
Pt	6.518	7.893	0.673	58205	12129
Pt^{2+}	5.483	4.257	0.618	160833	6541
Pt^{4+}	4.860	3.508	0.539	534241	5391
Au^{1+}	6.989	7.900	0.595	14048	12139
Au^{3+}	4.910	3.325	0.679	659142	5109
Hg^{1+}	5.370	4.935	1.360	103996	7583
Hg^{2+}	4.809	4.304	0.959	298500	6614

The presented histograms show for the most part a different influence of the central monatomic entity on the crystal structure in its closest environs as they do in the array of greater interatomic distances. In Fig. 1.2.4 the point diagram of values m_2 vs. m_1 is depicted. Pairs of estimates (m_1, m_2) may be considered as realization of the random vector of variables (X, Y) with a bivariant normal probability model and the density function of

$$f(x, y) = \frac{1}{2\pi\sigma_x\sigma_y \sqrt{1 - \varrho_{xy}^2}}$$

$$\times \exp\left\{\frac{1}{2(1 - \varrho_{xy}^2)}\left[\frac{(x - \mu_x)^2}{\sigma_x^2} - \frac{2\varrho_{xy}(x - \mu_x)(y - \mu_y)}{\sigma_x\sigma_y} + \frac{(y - \mu_y)^2}{\sigma_y^2}\right]\right\}$$

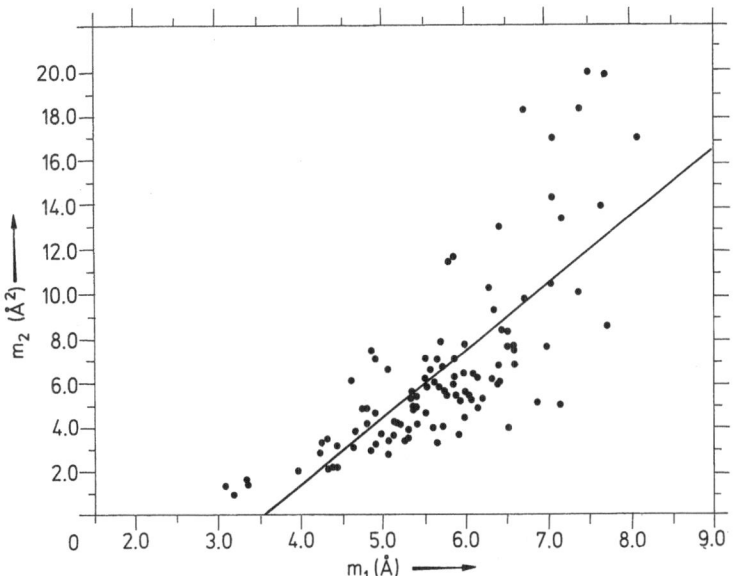

Fig. 1.2.4. Correlation of empirical estimates m_2 vs. m_1. The equation of regression line is $m_2 = -10.97 + (3.10)\, m_1$.

where μ_x, μ_y, σ_x and σ_y are expected values and the standard deviations respectively of the corresponding variables. ϱ_{xy} is their correlation coefficient introduced by (1.1.10). Estimation of the correlation coefficient is the *sample correlation coefficient*:

$$R_{XY} = \frac{1}{n-1} \sum_{i=1}^{n} \left(\frac{X_i - \bar{X}}{S_X}\right) \left(\frac{Y_i - \bar{Y}}{S_Y}\right)$$

S_x and S_y are sample standard deviations of variables X and Y. For concrete pairs of values (x_1, y_1), (x_2, y_2), (x_3, y_3), ... , (x_n, y_n) the *observed sample correlation coefficient* is used:

$$r_{xy} = \frac{1}{n-1} \sum_{i=1}^{n} \left(\frac{x_i - \bar{x}}{s_x}\right) \left(\frac{y_i - \bar{y}}{s_y}\right)$$

However, the modified formula is used:

$$r_{xy} = \frac{\sum_{i=1}^{n} x_i y_i - n\bar{x}\bar{y}}{\sqrt{\left(\sum_{i=1}^{n} x_i^2 - n\bar{x}^2\right)\left(\sum_{i=1}^{n} y_i^2 - n\bar{y}^2\right)}} \tag{1.2.3}$$

When $r_{xy} = 0$ variables X and Y are considered completely unrelated. If these values are equal to $+1$ or -1, variables X and Y are entirely related. In our case of correlation m_2 vs. m_1 is $r = 0.75$. Thus there is a significance in testing the hypothesis:

$$H_0 : r = 0 \quad \text{against} \quad H_1 : r \neq 0$$

For such a test we use the characteristic

$$T_{n-2} = \frac{r_{xy}\sqrt{n-2}}{\sqrt{1-r_{xy}^2}}$$

showing t distribution with $n - 2$ degrees of freedom. The tested quantity for $n = 107$ has the value of $T_{n-2} = 11.62$. On the significance level $\alpha = 0.05$ is $t_{n-2, 1-\alpha/2} = 1.97$ [44], i.e. $T_{n-2} > t_{n-2, 1-\alpha/2}$. The hypothesis H_0 of the mutual independence of m_1 and m_2 may be rejected. When we suppose a linear regression function

$$m_2 = a + bm_1 \qquad (1.2.4)$$

parameters a, b calculated by the *least squares method* based on the presumtion:

$$\sum_{i=1}^{n} [y_i - (a + bx_i)]^2 = \text{minimum}$$

are -10.97 Å^2 and 3.10 Å. Let us make the transformation of the random variable X to variable W:

$$W = X - a_0$$

where $a_0 \neq 0$ is a constant. Then the expected value of the transformed variable is according to (1.1.2)

$$\mu_W = E(W) = E(X - a_0) = E(X) - E(a_0) = \mu_X - a_0$$

For the empirical estimate of the expected value of variable, the relation may be written:

$$m_{1W} = m_1 - a_0 \qquad (1.2.5)$$

The variance of variable W can be expressed using formulae (1.1.2) and (1.1.4)

$$\begin{aligned}
\text{Var}(W) = \sigma_W^2 &= E[(W - \mu_W)^2] = E(W^2) - \mu_W^2 = \\
&= E[(X - a_0)^2] - (\mu_X - a_0)^2 = E[X^2 - 2a_0 X + a_0^2] - \\
&\quad - \mu_X^2 + 2\mu_X a_0 - a_0^2 = E(X^2) - 2a_0\mu_X + a_0^2 - \\
&\quad - \mu_X^2 + 2\mu_X a_0 - a_0^2 = E(X^2) - \mu_X^2 = \text{Var}(X) = \sigma_X^2
\end{aligned}$$

Thus for the variance of variable W the following applies:

$$m_{2W} = m_2 \qquad (1.2.6)$$

Substituting for m_1 and m_2 from (1.2.5) and (1.2.6) into (1.2.4) we obtain

$$m_{2W} = a + b(m_{1W} + a_0) = (a + ba_0) + bm_{1W}$$

We thus obtained a linear dependence between the first and the second central moments of variable W. When we further postulate that the second moment should be proportional to the first, then

$$a + ba_0 = 0, \quad \text{i.e.} \quad a_0 = -\frac{a}{b} = 3.55 \text{ Å}$$

Then $m_{2W} = bm_{1W}$ (Fig. 1.2.4), or

$$\sigma_W^2 = b\mu_W \tag{1.2.7}$$

For these moments in the case of random variable with Γ distribution the relations hold (Table 1.1.1): $\mu = \tau/\lambda$ and $\sigma^2 = \tau/\lambda^2$ or $\sigma^2 = \mu/\lambda$. Comparing the last relation with (1.2.7) we obtain $b = 1/\lambda = 3.10$ Å. Thus for variable W the Γ probability model with parameter $\lambda = \dfrac{1}{b} = 0.32 \text{ Å}^{-1}$ may be assumed. For the lengths of inter-atomic vectors with minimum value of 3.55 Å Γ probability model can be assumed.

Glossary of Symbols

$\varrho(x, y, z)$	Electron density
$F(hkl)$	Structure factor
G	Particle group
$\Sigma_0, \dots, \Sigma_{III}$	Symmetry class of particle group
d	Interatomic distance
r_A, r_B	Atomic radii
x_i	Interatomic distance in the unique set of crystallochemical unit
A_i	Point of sample space (event)
S	Sample space
w_i	Statistical weight
Pr	Probability
$\cup$	Symbol of union sets
$\cap$	Symbol of intersection sets
$\vert$	Symbol of conditioned events
ξ, X	Continuous random variables
$P(x)$	Probability distribution (distribution function) of random variable
$f(x)$	Density function or distribution density
μ	Expected value of random variable
$\sigma^2, \text{Var}(X)$	Variance of random variable
μ_k	k^{th} moment of random variable
ν_k	k^{th} central moment of random variable
ν	Variance coefficient
γ	Asymmetry of distribution
$\bar{X}$	Sample mean
S^2	Sample variance
$\bar{x}$	Observed sample mean
s^2	Observed sample variance

Z	Random variable exhibiting standard normal distribution
z_α	Critical value of variable Z
H	Statistical hypothesis
θ	Parameter of density function (distribution)
α	Significance level
β	Test power
M_α	Critical area
$\in$	Symbol expressing that given element belongs to the given set
n_{crit}	The critical sample size
T	Variable exhibiting t distribution
$t_{n-1;1-\alpha/2}$	Critical value of variable T
F	Variable exhibiting F distribution
Q_i	Quadratic form of random variables
$\text{Cov}(X)$	Covariance of random variable
$\varrho_{\xi,\eta}$	Correlation coefficient of random variables
f_ω	Density function of random vector
r	Interatomic vector; radius vector
$\mathbb{A}$	Geometric center of influence domain; common origin of interatomic vectors
Δ	Length of the data class
h_i	Middle of the data class
f_i	Frequence of occurence
m_i	Empirical estimate of the ith central moment
R_{XY}	Sample correlation coefficient
r_{xy}	Observed sample correlation coefficient
a, b	Coefficients of regression function
W	Transformed random variable

1.3 References

1. Schoenflies, A.: Kristallsysteme und Kristallstruktur, Leipzig 1891
2. Jørgensen, Ch. K.: Int. Rev. Phys. Chem. **1**, 225 (1981)
3. Weissenberg, K.: Z. Kristallogr. **62**, 52 (1925)
4. Valach, F., Kohout, J., Dunaj-Jurčo, M., Hvastijová, M., Gažo, J.: J. Chem. Soc. Dalton 1868 (1979)
5. Tideswell, N. W., Kruse, F. H., Mc Cullough, D. J.: Acta Crystallogr. **10**, 99 (1957)
6. Furuseth, S., Selte, K., Kjekshus, A.: Acta Chem. Scand. **19**, 257 (1965)
7. Wickoff, R. W. G.: Crystal Structures, Vol. 1, p. 110, New York—London—Sydney, Interscience Publ. John Wiley & Sons 1963²
8. Pauling, L.: Phys. Rev. **54**, 899 (1938)
9. Mott, N. F., Jones, N.: The Theory of the Properties of Metals and Alloys, Oxford, Oxford University Press 1936
10. Pauling, L.: The Nature of the Chemical Bond, Ithaca Cornell University Press 1960
11. Goldschmidt, V. M.: Geometrische Verteilungsgesetze der Elemente, Oslo 1926
12. Pauling, L.: Proc. Natl. Acad. Sci. **18**, 293 (1932)
13. Pauling, L., Brockway, L. O., Beach, J. Y.: J. Am. Chem. Soc. **57**, 2705 (1935)
14. Pauling, L.: J. Am. Chem. Soc. **69**, 542 (1947)
15. Shannon, R. D., Prewitt, C. D.: Acta Crystallogr., Sect. B, **25**, 925 (1969)
16. Peterson, J. R., Cunningham, B. B.: Inorg. Nucl. Chem. Lett. **3**, 327 (1967)
17. Peterson, J. R., Cunningham, B. B.: J. Inorg. Nucl. Chem. **30**, 1775 (1968)

18. Kálmán, A.: J. Chem. Soc., Chem. Commun. 1857 (1971)
19. Seitz, F.: The Modern Theory of Solids, New York, Mc Craw-Hill Book Co. 1940
20. Fukunaga, O., Fujita, T.: J. Solid State Chem. **8**, 331 (1973)
21. Greis, O., Petzel, T.: Z. Anorg. Allgem. Chem. **403**, 1 (1974)
22. Knop, O., Carlow, J. S.: Can. J. Chem. **52**, 2175 (1974)
23. Wolfe, R. W., Newnham, R. E.: J. Electrochem. Soc. **116**, 832 (1969)
24. Mc Carthy, G. J.: Mater. Res. Bull. **6**, 31 (1971)
25. Shannon, R. D.: Acta Crystallogr., Sect. A, **32**, 751 (1976)
26. Silva, R. J., Mc Dowell, Keller, W. J., Tarrant, J. R.: Inorg. Chem. **13**, 2233 (1974)
27. Allen, F. H., Bellard, S., Brice, M. D., Cartwright, B. A., Doubleday, A., Higgs, H.,
    Hummelink, T., Hummelink-Peters, B. G., Kennard, O., Motherwell, W. D. S., Rodgers,
    J. R. & Watson, D. G.: Acta Crystallogr., Sect. B, **35**, 2331 (1979)
28. Kennard, O., Watson, D. G., Town, W. G.: J. Chem. Doc. **13**, 14 (1973)
29. Allen, F. H., Kennard, O., Motherwell, W. D. S., Town, W. G. & Watson, A. D.: J. Chem.
    Doc. **13**, 119 (1973)
30. Allen, F. H., Kennard, O., Motherwell, W. D. S., Town, W. G., Watson, D. G.,
    Scott, T. Y. & Larson, A. C.: J. Appl. Crystallogr. **7**, 73 (1974)
31. Allen, F. H.: Acta Crystallogr., Sect. B, **37**, 890 (1981)
32. Allen, F. H., Isaacs, N. W., Kennard, O., Motherwell, E. D. S., Petersen, R. C.,
    Town, W. G. & Watson, D. G.: J. Chem. Doc. **13**, 211 (1973)
33. Eisen, M.: Introduction to Mathematical Probability Theory, New Jersey, Prentice-Hall
    1969
34. Feller, W.: An Introduction to Probability Theory and its Applications, Vol. 1, New York,
    Wiley 1950, Vol. 2, New York, Wiley 1966
35. Hennequin, P., Tortrat, A.: Théorie des probeblités et quel ques applications, Paris, Mason
    1965
36. Cramér, H.: Mathematical Methods of Statistics, Princeton 1946
37. Rao, R. C.: Linear Statistical Inference and its Applications, New York, Wiley 1973
38. Wilks, S.: Mathematical Statistics, New York, Wiley 1962
39. Guenther, W. C.: Concepts of Statistical Inference, Tokyo, Düsseldorf, Johannesburg,
    London, Mexico, New Delhi, Panama, Rio de Janeiro, Singapore, Sydney, International
    Student Edition, Mc Graw-Hill, Kogakusha, Ltd. 1973[2]
40. Kendall, M. G.: The Advenced Theory of Statistics, Vol. 1, London, Charles Griffin
    & Co Ltd. 1951
41. Richter, H.: Wahrscheinlichkeitstheorie, Berlin, Göttingen, Heidelberg 1956
42. Warden, B. L.: Mathematische Statistik, Berlin, Göttingen, Heidelberg 1957
43. Lehman, E. L.: Testing Statistical Hypothesis, New York, Wiley 1959
44. Hald, A.: Statistical Tables and Formulas, New York, John Wiley & Sons, Inc. 1952
45. Murray-Rust, P., Raftery, J.: J. Mol. Graphics **3**, 50 (1985)
46. Doms, L., Van Hemelrijk, D., Van de Mieroop, W., Lenstra, A. T. H. & Geise, H. J.:
    Acta Crystallogr., Sect. B, **41**, 270 (1985)
47. International Tables for Crystallography, Vol. A, Space-Group Symmetry, Dordrecht,
    Boston, D. Reidel Publishing Comp. 1983
48. Delaunay, B. N., Galinlin, R. V., Dolbin, R. V., Zalgaller, V. A. & Stogrin, M. I.:
    Dokl. Akad. Nauk SSSR **209**, 309 (1973)
49. Delaunay, B. N.: Z. Kristallogr. **84**, 109 (1933)
50. Nowacki, W.: Z. Kristallogr. **85**, 331 (1933)
51. Delone, B.: Acta Crystallogr., Sect. A, **21**, 28 (1966)
52. Minkowski, H.: Algemeine Lehrsätze über die konvexen Polyeder. In Gesammelte Ab-
    handlungen, Leipzig 1911 Chelsea Reprint 1967
53. Grünbaum, B.: Convex Polytopes, New York, Interscience 1967
54. Grünbaum, B., Shephard, G. C.: Bull. London Math. Soc. **1**, 257 (1969)
55. Jucovič, E.: Konvexné mnohosteny, Bratislava, Veda 1981
56. Delone, B. N., Dolbilin, N. P., Štogrin, M. J., Galjulin, M. J.: Dokl. Akad. Nauk SSSR
    227, 19 (1976)
57. Loeb, A. L.: J. Solid State Chem. **1**, 237 (1970)

58. Laves, F.: Crystal structure and atomic size, in: Theory of Alloys, Cleveland, Ohio, Amer. Soc. Met. 1956
59. Pearson, W. B.: A Handbook of Lattice Spacings and Structures of Metals and Alloys 2, p. 1, Oxford, Pergamon Press 1967
60. Valach, F., Koreň, B., Sivý, P. & Melník, M.: Structure and Bonding **55**, 103 (1983)
61. Cambridge Crystallographic Data Centre (1984), University Chemical Laboratory, Cambridge, England
62. SIEMENS PBS 4004, MEB METHODENBANK 1973

2 Factors Influencing the Dispersion of Lengths of Interatomic Vectors

Table 2.1 presents empirical estimates of the first two central moments of inter-atomic vector lengths with the described properties. In this case terminal monatomic entities belong to a certain type. The dispersion of lengths of these vectors expressed by the second central moment m_2 shows the value of 7.383 Å^2 for the sample of vector lengths of $Zn^{2+} \rightarrow 0$, this value being much higher than 5.192 Å^2 for the sample of vector lengths for $Zn^{2+} \rightarrow S$. In the case of such great sample sizes the variance equality of these data sets can be tested using sample characteristics of $F = S_1^2/S_2^2$:

$$H_0 : \sigma_1^2 = \sigma_2^2 \text{ against } H_1 : \sigma_1^2 \neq \sigma_2^2$$

Table 2.1. Empirical estimates of the first and the second central moments (m_1 and m_2) and sample sizes (n) of interatomic vector lengths for some types of central monatomic entities. Terminal entities of these vectors are of a certain type. Critical values of sample sizes (n_{crit}) were calculated by formula (1.1.8); $\alpha = 0.05$

Set of interatomic vectors	m_1 [Å]	m_2 [Å^2]	n	n_{crit}
$Co^{3+} \rightarrow N$	5.300	8.091	54528	12433
$Co^{3+} \rightarrow Cl$	6.044	6.896	58340	10597
$Zn^{2+} \rightarrow N$	7.204	5.053	69141	7747
$Zn^{2+} \rightarrow O$	6.300	7.383	81831	11345
$Zn^{2+} \rightarrow S$	7.270	5.192	91041	7978
$Cu^{2+} \rightarrow N$	5.038	8.177	40861	12565
$Cu^{2+} \rightarrow Cl$	4.933	8.647	25251	13287

We will use the estimate of the above characteristics $F_1 = m_2(Zn^{2+} \rightarrow 0)/m_2(Zn^{2+} \rightarrow S)$ = 1.422. The critical F value on the significance level of $\alpha = 0.05$ for $v_1 = n(Zn^{2+} \rightarrow 0)$ $- 1$ and $v_2 = n(Zn^{2+} \rightarrow S) - 1$ is 1.011 [1]. Since $F > F_{crit}$, the hypothesis H_0 of the equality of dispersions can be rejected. The vector lengths of $Co^{3+} \rightarrow Cl$ and $Co^{3+} \rightarrow N$ also show a significant difference between their dispersions. In both cases of central monatomic entities one could explain the differences between the dispersions of interatomic vector lengths with different qualities of their terminal entities. The values of m_2 of the samples of vectors of $Zn^{2+} \rightarrow S$ and $Zn^{2+} \rightarrow N$ exhibit, however, only a small difference (0.139 Å^2). Similarly, the estimates of the

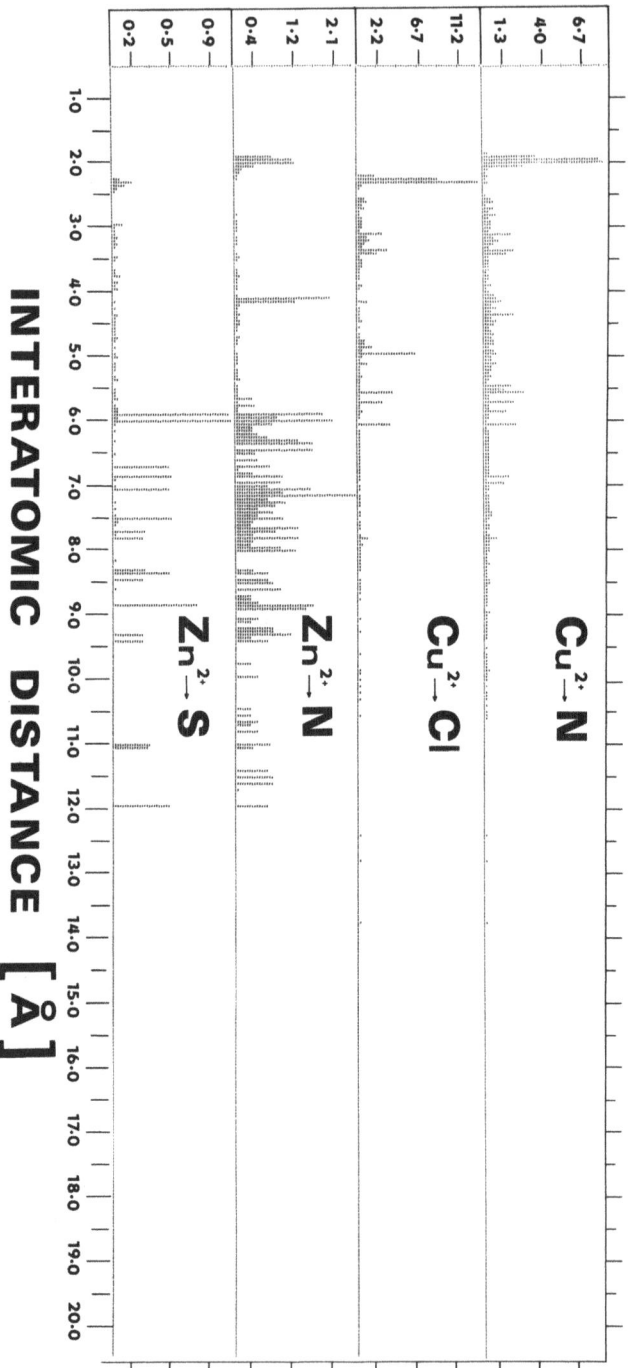

second central moments of samples $Cu^{2+} \rightarrow Cl$ and $Cu^{2+} \rightarrow N$ are comparatively near.

The histograms of lengths of these vectors show, however, rather large differences in the values of relative frequencies. In these cases the F values make 1.028 and 1.058. The corresponding critical values for $\alpha = 0.05$ are 1.012 and 1.015. Also in these cases $F > F_{crit}$ is again explainable with a different quality of terminal atoms. The difference between F and F_{crit} is here, however, substantially lower ($F \approx F_{crit}$). So it appears to be useful to test the equality $\sigma_1^2 = k\sigma_2^2$, k being a number near to one. Thus we formulate the hypothesis

$$H_0 : \sigma_1^2/\sigma_2^2 = k \text{ against } H_1 : \sigma_1^2/\sigma_2^2 \neq k$$

In this case of two sided hypothesis H_1 ($\sigma_1^2/\sigma_2^2 > k$, or $\sigma_1^2/\sigma_2^2 < k$), H can be accepted only when the inequality

$$\Pr\left[k/F_{\alpha, v_1, v_2} < \frac{\sigma_1^2}{\sigma_2^2} < kF_{\alpha, v_1, v_2} \right] = 1 - 2\alpha \qquad (2.1)$$

is fulfilled. If we take $k = 1.03$ for samples $Zn^{2+} \rightarrow S$, $Zn^{2+} \rightarrow N$ and $Cu^{2+} \rightarrow Cl$, $Cu^{2+} \rightarrow N$ the relation $k/F_{crit} < F < kF_{crit}$ holds. The probability (2.1) is $\Pr = 1 - 2\alpha = 90\%$. That means that in a hundred interatomic vectors, in one set, or other, there will be 90 of the same kind, for which it holds: $\sigma_1^2 = (1.03)\,\sigma_2^2$.
Therefore it may be assumed:

(a) There exist at least two factors influencing the dispersion of interatomic vector lengths; the factor connected with the properties of the central monatomic entity, and the factor connected with the variety of terminal atom types.
(b) We can take such samples of interatomic vector lengths of the normal type of central monatomic entity, the dispersion of which depends on the central entity type, i.e. it is a characteristical quantity of the given central monatomic entity with a certain inaccuracy (tolerance).

The conclusion of (b) leads to the idea of a certain factor averaging, caused by the variability of crystal structures as wholes.

From the classical thermodynamic point of view, it is well known that an isolated system can pass into the state with a maximum possible enthropy. As a matter of fact it is a contention of the second thermodynamic law (dS > 0). If we do have a set of interatomic vectors with a certain type of central monatomic entity and arbitrary type of terminal entities, the size of an uncertainty of such a system, or entropy, according to Shanon [2, 3] is

$$S = \sum_i p_i \ln p_i$$

For interatomic vector lengths exists a probability of their occurrence in i^{th} class p_i. Deriving from the given formula for enthropy from mostly physical point of view has been discussed [4, 5] and for this reason will not be here. Such entropy has the following properties:

(i)   is positive
(ii)   increases with increasing uncertainty
(iii) is additive for an independent source of uncertainty.

Predominance of maximum entropy describes our degree of knowledge about the system. In the case to two mutually independent factors given in (a), a multiplicative form of probabilities $p_i = u_i v_i$ may be proposed. Let the factor connected with the properties of central monatomic entity be represented here by probabilities u_i and the factor due to variability of structures be represented by probabilities v_i. The second factor is manifested mainly in the vectors whose lengths are greater then a_0 (Fig. 1.2.4). Since in this region Γ distribution may be proposed, we may expect that maximum entropy will lead to uniform distribution of probabilities (v_i = const.) with zero variance. That way one can propose a confidence interval in the framework, in which distribution of probabilities v_i is uniform.

From the view of the quantum theory of solids the electron moving in the crystal field may be described by the wave function $\psi(r)$, where r is the radius vector of the electron which can be expressed in the crystallographic coordinate system with the unit cell vectors a, b, c:

$$r = xa + yb + zc$$

For this function the Schrödinger equation holds:

$$\mathbb{H}(r)\,\psi(r) = E\psi(r) \tag{2.2}$$

where Hamiltonian can be expressed as

$$\mathbb{H}(r) = -\frac{\hbar^2}{2m}\Delta + U(r) = -\frac{\hbar^2}{2m}\Delta - eV(r)$$

$U(r)$ is the potential energy and $V(r)$ is the threedimensional periodic potential of the crystal field. Thus, if the lattice vector is expressed in the form of

$$r = p_1 a + p_2 b + p_3 c$$

where p_1, p_2, p_3 are arbitrary integers for the crystal field potential, the relation must apply $V(r + r_p) = V(r)$ and likewise for Hamiltonian the relation: $\mathbb{H}(r + r_p) = \mathbb{H}(r)$[1] will be valid. Schrödinger Eq. (2.2) may then be also written in the form of

$$\mathbb{H}(r)\,\psi(r + r_p) = E\psi(r + r_p)$$

This equation is fulfilled when the electron wave function, after changing its position by r_p, will differ by the constant multiplication factor

$$\psi(r + r_p) = k_p \psi(r)$$

---

[1] See Chap. 3.4

Then for basic translations the relations hold:

$$\psi(r + a) = K_1\psi(r)$$

$$\psi(r + b) = K_2\psi(r)$$

$$\psi(r + c) = K_3\psi(r)$$

The multiplication constants here are generally complex numbers, thus they may be expressed in the form of

$$K_1 = e^{ik_1}, \; K_2 = e^{ik_2}, \; K_3 = e^{ik_3}$$

where K_1, K_2, K_3 are real numbers. Similarly for integer multiplications to the basic translations, the relations will hold:

$$\psi(r + p_1 a) = e^{ip_1 k_1}\psi(r)$$

$$\psi(r + p_2 b) = e^{ip_2 k_2}\psi(r)$$

$$\psi(r + p_3 c) = e^{ip_3 k_3}\psi(r)$$

By connecting these functions we obtain the resulting vawe function

$$\psi(r + r_p) = e^{i(p_1 k_1 + p_2 k_2 + p_3 k_3)}\,\psi(r) \tag{2.3}$$

Then let us introduce the vector

$$k^* = k_1 a^* + k_2 b^* + k_3 c^*$$

where a^*, b^*, c^* are unit cell vectors of the *reciprocal lattice* defined by the relations

$$a^* = \frac{b \times c}{a.(b \times c)} \qquad b^* = \frac{c \times a}{a.(b \times c)} \qquad c^* = \frac{a \times b}{a.(b \times c)}$$

The validity of the relations

$$
\begin{array}{lll}
a.a^* = 1 & a.b^* = 0 & b.c^* = 0 \\
b.b^* = 1 & a.c^* = 0 & c.a^* = 0 \\
c.c^* = 1 & b.a^* = 0 & c.b^* = 0
\end{array}
$$

may be easily proved and by their mutual multiplication we shall obtained the identity

$$r_p \cdot k^* = p_1 k_1 + p_2 k_2 + p_3 k_3$$

Inserting this result into Eq. (2.3) will give

$$\psi(r + r_p) = e^{i(k^* \cdot r_p)}\,\psi(r)$$

This equation allows us to express

$$\psi(\mathbf{r}) = e^{-i(\mathbf{k}^* \cdot \mathbf{r}_p)} \psi(\mathbf{r} + \mathbf{r}_p) = e^{-i\mathbf{k}^* \cdot (\mathbf{r}+\mathbf{r}_p)} \psi(\mathbf{r} + \mathbf{r}_p) e^{i\mathbf{k}^* \cdot \mathbf{r}}$$

Introducing the notation of the function

$$u(\mathbf{k}^*, \mathbf{r}) = e^{-i\mathbf{k}^* \cdot (\mathbf{r}+\mathbf{r}_p)} \psi(\mathbf{r} + \mathbf{r}_p) \qquad (2.4)$$

we obtain the vawe function

$$\psi(\mathbf{k}, \mathbf{r}) = u(\mathbf{k}, \mathbf{r}) e^{i\mathbf{k} \cdot \mathbf{r}}$$

known as the Bloch function [6, 7].

Function (2.4) again is periodical with periods $\mathbf{r}_p$

$$u(\mathbf{k}, \mathbf{r} + \mathbf{r}_p) = u(\mathbf{k}, \mathbf{r})$$

In the periodical crystal field, i.e. in describing the electron state the period of the field potential always appears.

Wanting, individually, to describe the atom properties in the crystal it becomes necessary to formally replace the periodical field potential of the crystal with the average constant potential. Let us therefore consider n potentials of mutually independent crystal fields ($V_1, V_2, V_3 ... V_n$). Each of these potentials can be expressed as the sum of an infinite Fourier development (Appendix C):

$$V_i(\mathbf{r}) = \sum_{g_1} \sum_{g_2} \sum_{g_3} \varphi_i(g_1 g_2 g_3) e^{2\pi i \mathbf{g}^*(g_1 g_2 g_3) \cdot \mathbf{r}}$$

$\mathbf{r}$ is the radius vector in the Cartesian coordinate system with the coordinates x_c, y_c, z_c. The origin of this system will be placed in the centre of the considered monatomic entity.

$$\mathbf{r} = x_c \mathbf{i} + y_c \mathbf{j} + z_c \mathbf{k}$$

$\mathbf{g}^*$ is the reciprocal lattice vector also in this rectangular coordinate system:

$$\mathbf{g}^* = g_1 \mathbf{a}^* + g_2 \mathbf{b}^* + g_3 \mathbf{c}^*$$

with g_1, g_2, g_3 being integers. The coefficients of this development may be expressed by the formula:

$$\varphi_i(g_1 g_2 g_3) = \frac{1}{\Omega_{0i}} \int_{(\Omega_{0i})} V_i(\mathbf{r}) e^{-2\pi i \mathbf{g}^*(g_1 g_2 g_3) \cdot \mathbf{r}} d\Omega \qquad (2.5)$$

where Ω_{0i} is the unit cell volume of ith crystal and $d\Omega = dx_c\, dy_c\, dz_c$. The total potential of the cumulative field of n crystal fields is then

$$V(\mathbf{r}) = \sum_{i=1}^{n} V_i(\mathbf{r}) = \sum_{i}\sum_{g_1}\sum_{g_2}\sum_{g_3} \varphi_i(g_1 g_2 g_3)\, e^{2\pi i \mathbf{g}^*(g_1 g_2 g_3)\cdot \mathbf{r}}$$

Using Eq. (2.5) one coefficient of this development can be expressed as

$$\varphi_i(000) = \frac{1}{\Omega_{0i}} \int\limits_{(\Omega_{0i})} V_i(\mathbf{r})\, d\Omega$$

this being the average value of ith potential.

The cumulative potential is:

$$V(\mathbf{r}) = \sum_{i=1}^{n} \varphi_i(000) + \sum_{i=1}^{n}\sum_{g_1}\sum_{g_2}\sum_{g_3} \varphi_i(g_1 g_2 g_3)\, e^{2\pi i \mathbf{g}^*(g_1 g_2 g_3)\cdot \mathbf{r}}$$

$$= \sum_{i=1}^{n} \varphi_i(000) + \sum_{i=1}^{n}\sum_{g_1}\sum_{g_2}\sum_{g_3} \varphi_i(g_1 g_2 g_3)\, e^{2\pi i (g_1 x_c + g_2 y_c + g_3 z_c)}$$

$$= V_{0n} + V_n'(x_c, y_c, z_c)$$

Then there is non-zero probability of selection such an infinite set of crystal fields with potentials $V_1(\mathbf{r})$, $V_2(\mathbf{r})$, $V_3(\mathbf{r})$ ... , for which the second component of the cumulative potential is equal to zero.

2.1 Application of Variance Analysis

The used data base allows crystal structure data of substances to be attained, between which a certain dependence and repeated data could occur, thus introducing inaccuracy into calculation of the variance estimates as the measure of dispersion. Testing the position of terminal monatomic entity, whether it is contained in the influence domain of the central entity, is based on parameters of the unit cell. Its selection is in some cases a matter of convention. An unconventional selection of the unit cell can also lead to undesirable data repeating which will then be involved in the variance calculation. Therefore it is necessary to use such a technique, for the given data sample, that would enable us to make decisions on the significance also of these inaccuracies. It may be expected that their contribution will not be significant in cases of random samples of interatomic vector lengths from a greater quantity of crystal structures. If the terminal monatomic entities are of different types in a randomly chosen set of crystal structures, then the dispersion of interatomic vector lengths is caused by properties of the central entity and by the diverse nature of atoms contained in the domains of influence. There are at least two factors influencing the dispersion of frequencies: the factor given by the central entity properties and the factor determined by the properties of the other components of the crystal structure. Further on, we shall suppose a mutual independence of these factors. Now the problem arises of how to eliminate the influence of the crystal structure factor, so that the dispersion of distances should reflect the central entity properties only. It is the *variance analysis*, that verifies the effect of one or more factors on the level of the quantitative feature under investigation. Originally this method was developed to evaluate experiments in biology. At present, however, it is still being applied in different branches of natural sciences and it represents one of the most general statistical methods. With respect to statistical mathematics it was worked out for the first time by R. A. Fisher in his monograph, The Design of Experiment (1925). A suitable explanation of variance analysis is also found in [6–13].

In crystal structures there are different factors, effectively called in literature "steric effect", "solid state effect" and others. Variance analysis gives the possibility of quantitative valuation of such factors on the dispersion of the chosen random variable. In applying this method we shall be interested in a certain averaging of factors, causing also extrinsic distortions of the central entity inner coordination sphere, and we will call them *extrinsic factors*. The factor given by properties of the central monatomic entity will be the *intrinsic factor*. Introducing a test for the significance of averaging extrinsic factors, allows us to assign to m_2 quantities of a more general signification for the given central entity with K electron. Calculation of empirical distribution of interatomic vector lengths for the central monatomic entities shows in most examples their complexity and high asymmetry (Table 1.1.3). For another statistical treatment from this point of view it will be more suitable to use frequencies of interatomic lengths.

Let us consider n central monatomic entities of the same type from crystallochemical units[1] of s ($s \leq n$) randomly elected crystal structures. Then by calculation we can obtain n radial distributions of interatomic vectors. If x_{ij} signs the frequency of interatomic vector lengths of ith central entity being in the jth class then the matrix can be construed:

$$
\begin{bmatrix}
x_{11} & x_{21} & x_{31} & \cdots & x_{n1} \\
x_{12} & x_{22} & x_{32} & \cdots & x_{n2} \\
x_{13} & x_{23} & x_{33} & \cdots & x_{n3} \\
\vdots & \vdots & \vdots & & \vdots \\
x_{1m} & x_{2m} & x_{3m} & \cdots & x_{nm}
\end{bmatrix}
\tag{2.1.1}
$$

Here, for simplicity's sake, we assume the same number of classes, m. This is then an *orthogonal case*. Regarding the above mentioned factors it appears reasonable to presume

$$
x_{ij} = \mathscr{I}_{ij} + \mathscr{E}_i
\tag{2.1.2}
$$

The random variable $\mathscr{I}_{ij}$ represents the intrinsic, while $\mathscr{E}_i$ the contribution of extrinsic factor. At a random election of crystal structures the second factor has a random effect and therefore $\mathscr{E}_i$ is also a random variable and thus the model expressed by

---

[1] Unique set of residues in structure.

Eq. (2.1.2) is called 2nd *mathematical model with random effect*. The mean value of variable x_{ij} is

$$E(x_{ij}) = \sum_{i=1}^{n} \sum_{j=1}^{m} p_{ij} x_{ij}$$

where p_{ij} is the probability of x_{ij}. Then the variance of this variable is

$$\text{Var}\,(x_{ij}) = \sigma_x^2 = \sum_{i=1}^{n} \sum_{j=1}^{m} p_{ij}[x_{ij} - E(x_{ij})]^2$$

$$= \sum_{i=1}^{n} \sum_{j=1}^{m} p_{ij}[x_{ij}^2 - 2x_{ij}E(x_{ij}) + E^2(x_{ij})]^2$$

$$= \sum_{i=1}^{n} \sum_{j=1}^{m} p_{ij}x_{ij}^2 - 2E(x_{ij}) \sum_{i=1}^{n} \sum_{j=1}^{m} p_{ij}x_{ij} + E^2(x_{ij}) \sum_{i=1}^{n} \sum_{j=1}^{m} p_{ij}$$

$$= E(x_{ij}^2) - 2E^2(x_{ij}) + E^2(x_{ij}) = E(x_{ij}^2) - E^2(x_{ij}) \tag{2.1.3}$$

because $\sum_{i=1}^{n} \sum_{j=1}^{m} = p_{ij} = 1$. Using this identity and model (2.1.2) the variance σ_x^2 may be further expressed as

$$\text{Var}\,(x_{ij}) = E[(\mathscr{I}_{ij} + \mathscr{E}_i)^2] - E^2(\mathscr{I}_{ij} + \mathscr{E}_i)$$

$$= E[\mathscr{I}_{ij}^2 + 2\mathscr{I}_{ij}\mathscr{E}_i + \mathscr{E}_i^2] - [E(\mathscr{I}_{ij}) + E(\mathscr{E}_i)]^2$$

$$= E(\mathscr{I}_{ij}^2) + 2E(\mathscr{I}_{ij}\mathscr{E}_i) + E(\mathscr{E}_i^2) - E^2(\mathscr{I}_{ij}) - 2E(\mathscr{I}_{ij})E(\mathscr{E}_i) - E^2(\mathscr{E}_i) \tag{2.1.4}$$

Since we assume independence of variables $\mathscr{I}_{ij}$ and $\mathscr{E}_{ij}$, the probability p_{ij} may be expressed as the product $p_{ij} = p_i \cdot p_j$. Then the identity holds:

$$E(\mathscr{I}_{ij}\mathscr{E}_i) = \sum_{i=1}^{n} \sum_{j=1}^{m} p_{ij}\mathscr{E}_i\mathscr{I}_{ij} = \sum_{i=1}^{m} \sum_{j=1}^{n} p_i p_j \mathscr{E}_i\mathscr{I}_{ij}$$

$$= \left(\sum_{i=1}^{n} p_i\mathscr{E}_i \right) \left(\sum_{i=1}^{n} \sum_{j=1}^{m} p_j\mathscr{I}_{ij} \right) = E(\mathscr{E}_i)\,E(\mathscr{I}_{ij}) \tag{2.1.5}$$

Inserting Rq. (2.1.5) into Eq. (2.1.4) and using identity (2.1.3) for variables $\mathscr{I}_{ij}$ and $\mathscr{E}_i$, we obtain

$$\text{Var}\,(x_{ij}) = \text{Var}\,(\mathscr{I}_{ij}) + \text{Var}\,(\mathscr{E}_i) \tag{2.1.6}$$

Without any detriment of its generality model (2.1.2) can also be written in the form:

$$x_{ij} = \mu + \mathscr{I}_{ij} + \mathscr{E}_i$$

where μ is the *general mean*

$$\mu = \frac{1}{mn} \sum_{i=1}^{n} \sum_{j=1}^{m} x_{ij}$$

The extrinsic factor effect for the case of ith central monatomic entity is expressed by the difference

$$\mathscr{E}_i = \mu_i - \mu = \frac{1}{m} \sum_{j=1}^{m} x_{ij} - \mu$$

If we wish to elucidate the physical significance of the variance $\mathscr{E}_i$, we have to accept the assumptions:

(a) the random variables $\mathscr{I}_{ij}$ and $\mathscr{E}_i$ do not mutually correlate, i.e. $E(\mathscr{I}_{ij}\mathscr{E}_i) = 0$ for all i-s and j-s;
(b) the variables $\mathscr{I}_{ij}$ ($i = 1, 2, 3, \ldots n$; $j = 1, 2, 3, \ldots m$) do not mutually correlate, i.e.
$$E(\mathscr{I}_{ij}\mathscr{I}_{i'j'}) = 0 \text{ for all } i\text{-s, } i'\text{-s, } j\text{-s and } j'\text{-s.}$$
(c) the variances of variables E_i for all i-s are the same, i.e. Var $(E_i) = \sigma_E^2$ for all i-s.
(d) the point estimates of variables $\mathscr{I}_{ij}$ and $\mathscr{E}_i$ are zero, i.e. $E(\mathscr{E}_i) = E(\mathscr{I}_{ij}) = 0$

In our case of going about the study of coordination compounds, intrinsic effect of central monatomic entity is usually expressed differently from extrinsic effect of more distant entities. Assumption (a) is then satisfactorily fulfilled. Owing to miscellaneous non-central entities comprised in influence domains, we may regard assumption (b) as being successfully fulfilled in the framework of each structure, as well as in the framework of all set structures with the studied central entity. Since in this set central monatomic entity is always of the same type, we will regard assumption (c) as fulfilled. Assumption (d) also appears to be acceptable after introduction into model (2.1.2) the general mean μ. Then the covariance between frequencies of different classes ($j \neq j'$) is

$$\begin{aligned}
\text{Cov } (x_{ij}, x_{ij'}) &= E\{[x_{ij} - E(x_{ij})] [x_{ij'} - E(x_{ij'})]\} \\
&= E[(\mu + \mathscr{E}_i + \mathscr{I}_{ij} - \mu)(\mu + \mathscr{E}_i + \mathscr{I}_{ij'} - \mu)] \quad (2.1.7) \\
&= E(\mathscr{E}_i^2) + E(\mathscr{I}_{ij}\mathscr{E}_i) + E(\mathscr{E}_i\mathscr{I}_{ij'}) + E(\mathscr{I}_{ij}\mathscr{I}_{ij'}) = \text{Var } (\mathscr{E}_i) = \sigma_{\mathscr{E}}^2
\end{aligned}$$

Thus variance $\sigma_{\mathscr{E}}^2$ is the measure of mutual covariance between the frequencies of different radial distributions of interatomic vectors in matrix (2.1.1). If their mutual correlation is weak, then the dispersion of interatomic vectors lengths owing to extrinsic factors and correlations of data, will be near zero. This means a certain averaging of extrinsic factor and inaccuracy minimisation caused by repetition or mutual correlation of frequency distributions. Thus the contribution of $\sigma_{\mathscr{E}}^2$ variance to the overall variance (2.1.6) will be negligible, when this correlation will be minimum. This can be achieved in the concrete case of a set of s structures by stereochemical and crystallostructural variety of environments of n central monatomic entities contained in their domains of influence, as by a sufficiently great n.

Further on we will deduce formulae for point estimation of components of the whole variance, let us call them $\delta_{\mathcal{I}}^2$ and $\delta_{\mathcal{E}}^2$. For the matrix of frequencies (2.1.1) we will use designations for partial sums as follows:

$$x_{i.} = \frac{1}{m} \sum_{j=1}^{m} x_{ij} \qquad x_{..} = \frac{1}{N} \sum_{i=1}^{n} \sum_{j=1}^{m} x_{ij} \qquad N = nm$$

then it holds $x_{i.} = \mu + \mathcal{E}_i + \mathcal{I}_{i.}$ where $\mathcal{I}_{i.} = \frac{1}{m} \sum_{j=1}^{m} \mathcal{I}_{ij}$ and $x_{ij} - x_{i.} = \mathcal{I}_{ij} - \mathcal{I}_{i.}$.

Then let us calculate the expected value of the sum

$$E \sum_{i=1}^{n} \sum_{j=1}^{m} (x_{ij} - x_{i.})^2 = \sum_{i=1}^{n} \sum_{j=1}^{m} E(x_{ij} - x_{i.})^2$$

However, the identity holds

$$(x_{ij} - x_{i.})^2 = (\mathcal{I}_{ij} - \mathcal{I}_{i.})^2 = \mathcal{I}_{ij}^2 - 2\mathcal{I}_{ij}\mathcal{I}_{i.} + \mathcal{I}_{i.}^2$$

and the corresponding expected value is

$$E(x_{ij} - x_{i.})^2 = E(\mathcal{I}_{ij}^2) - 2E(\mathcal{I}_{ij}\mathcal{I}_{i.}) + E(\mathcal{I}_{i.}^2)$$

From this it follows

$$E \sum_{i=1}^{n} \sum_{j=1}^{m} (x_{ij} - x_{i.})^2 = \sum_{i=1}^{n} \sum_{j=1}^{m} E(\mathcal{I}_{ij}^2) - 2 \sum_{i=1}^{n} \sum_{j=1}^{m} E(\mathcal{I}_{ij}\mathcal{I}_{i.}) + \sum_{i=1}^{n} \sum_{j=1}^{m} E(\mathcal{I}_{i.}^2)$$

$$= nm\sigma_{\mathcal{I}}^2 - 2 \sum_{i=1}^{n} \sum_{j=1}^{m} E\left(\mathcal{I}_{ij} \frac{1}{m} \sum_{j=1}^{m} \mathcal{I}_{ij}\right) + \sum_{i=1}^{n} \sum_{j=1}^{m} \frac{1}{m^2} E\left(\sum_{j=1}^{m} \mathcal{I}_{ij}\right)^2$$

$$= nm\sigma_{\mathcal{I}}^2 - 2nm \frac{1}{m} \sigma_{\mathcal{I}}^2 + n\sigma_{\mathcal{I}}^2 = n(m - 1)\sigma_{\mathcal{I}}^2$$

Thus the dispersion of frequencies of interatomic vectors as a consequence of intrinsic factor, called further on *intrinsic dispersion*, shows the variance

$$\sigma_{\mathcal{I}}^2 = \frac{E \sum_{i=1}^{n} \sum_{j=1}^{m} (x_{ij} - x_{i.})^2}{N - n} \tag{2.1.8}$$

Then the point estimate of intrinsic variance is

$$\delta_{\mathcal{I}}^2 = \sum_{i=1}^{n} \sum_{j=1}^{m} (x_{ij} - x_{i.})^2/(N - n) \tag{2.1.9}$$

A similar way is used to express the point estimate of extrinsic variance of interatomic vector lengths. For this purpose let us calculate the expected value of a variable:

$$E\left[m \sum_{i=1}^{n} (x_{i.} - x_{..})^2\right] = m \sum_{i=1}^{n} E(x_{i.} - x_{..})^2$$

where

$$x_{..} = \mu + \mathcal{E}_. + \mathcal{I}_{..}, \qquad \mathcal{E}_. = \frac{1}{n} \sum_{i=1}^{n} \mathcal{E}_i \quad \text{and}$$

$$\mathcal{I}_{..} = \frac{1}{mn} \sum_{i=1}^{n} \sum_{j=1}^{m} \mathcal{I}_{ij}$$

Then the identities hold

$$x_{i.} - x.. = (\mathcal{E}_i - \mathcal{E}_.) + (\mathcal{I}_{i.} - \mathcal{I}..)$$
$$(x_{i.} - x..)^2 = (\mathcal{E}_i - \mathcal{E}_.)^2 + (\mathcal{I}_{i.} - \mathcal{I}..)^2 + 2(\mathcal{E}_i - \mathcal{E}_.)(\mathcal{I}_{i.} - \mathcal{I}..)$$

Based on assumption (a) the mean value of this variable is

$$E(x_{i.} - x..)^2 = E(\mathcal{E}_i - \mathcal{E}_.)^2 + E(\mathcal{I}_{i.} - \mathcal{I}..)^2 + 0$$

In inspecting the other assumption one can calculate for the addends on the right side

$$E(\mathcal{E}_i - \mathcal{E}_.)^2 = E(\mathcal{E}_i^2) - 2E(\mathcal{E}_i\mathcal{E}_.) + E(\mathcal{E}_.^2) =$$

$$= \frac{1}{n} \sigma_{\mathcal{I}}^2 - \frac{2}{nm} \sigma_{\mathcal{I}}^2 + \frac{1}{nm} \sigma_{\mathcal{I}}^2 = \frac{m-1}{nm} \sigma_{\mathcal{I}}^2$$

$$E(\mathcal{I}_{i.} - \mathcal{I}_{..})^2 = E(\mathcal{I}_{i.})^2 - 2E(\mathcal{I}_{i.}\mathcal{I}_{..}) + E(\mathcal{I}^2)$$

$$= \frac{1}{n} \sigma_{\mathcal{I}}^2 - \frac{2}{nm} \sigma_{\mathcal{I}}^2 + \frac{1}{nm} \sigma_{\mathcal{I}}^2 = \frac{m-1}{nm} \sigma_{\mathcal{I}}^2$$

By introducing these results, we obtain

$$E\left[m \sum_{i=1}^{n} (x_{i.} - x_..)^2 \right] = m \sum_{i=1}^{n} \frac{n-1}{n} \left[\sigma_{\mathcal{E}}^2 + \frac{1}{n} \sigma_{\mathcal{I}}^2 \right] = (n-1)(\sigma_{\mathcal{I}}^2 + m\sigma_{\mathcal{E}}^2)$$

$$E\left[\sum_{i=1}^{n} (x_{i.} - x_..)^2/(n-1) \right] = \sigma_{\mathcal{I}}^2 + m\sigma_{\mathcal{E}}^2$$

On substituting result (2.1.8) for $\sigma_{\mathcal{I}}^2$

$$E\left[\sum_{i=1}^{n} (x_{i.} - x_..)^2/(n-1) \right] = E\left[\sum_{i=1}^{n} \sum_{j=1}^{m} (x_{ij} - x_{i.})^2/(N-n) \right] + m\sigma_{\mathcal{E}}^2$$

from which we can calculate the extrinsic variance

$$\sigma_{\mathcal{E}}^2 = E\left[\sum_{i=1}^{n} (x_{i.} - x_..)^2/(n-1) m - \sum_{i=1}^{n} \sum_{j=1}^{m} (x_{ij} - x_{i.})^2/(N-n) m \right]$$

Thus the point estimate of the extrinsic variance is

$$\delta_{\acute{e}}^2 = \sum_{i=1}^{n} (x_{i.} - x_{..})^2/(n-1)\, m - \sum_{i=1}^{n} \sum_{j=1}^{m} (x_{ij} - x_{i.})^2/(N-n)\, m \qquad (2.1.10)$$

Now we will deduce a testing criterion for the zero hypothesis

$$H_0 : \sigma^2 = 0 \text{ against } H_1 : \sigma^2 \neq 0$$

Such a criterion will allow us to determine whether averaging of extrinsic factors will be sufficient, if we accept error α. Point estimates $(\delta_{\mathscr{I}}^2 + m\delta_{\acute{e}}^2)\,(n-1)$ and $\delta_{\mathscr{I}}^2(N-n)$ have according to the Cochran theorem the distributions $\chi_{v_1}^2$ and $\chi_{v_2}^2$ where $v_1 = n - 1$ and $v_2 = N - n$. Let us search for the distribution of the ratio of these two random variables:

$$\frac{\delta_{\mathrm{I}}^2 + m\delta_{\mathrm{E}}^2}{\delta_{\mathrm{I}}^2} \qquad (2.1.11)$$

Both variables are mutually independent and thus the probability density of the random vector (ξ_1, ξ_2) where $\xi_1 = \delta_{\mathscr{I}}^2 + m\delta_{\acute{e}}^2$, $\xi_2 = \delta_{\mathscr{I}}^2$ is

$$f_{v_1, v_2}(y_1, y_2) = v_1 f_{\chi_{v_1}^2}(v_1 y_1)\, v_2 f_{\chi_{v_2}^2}(v_2 y_2)$$

$$= \frac{v_1^{\frac{v_1}{2}} v_2^{\frac{v_2}{2}}}{2^{\left(\frac{v_1 + v_2}{2}\right)} \Gamma\left(\frac{v_1}{2}\right) \Gamma\left(\frac{v_2}{2}\right)} y_1^{\left(\frac{v_1}{2}-1\right)} y_2^{\left(\frac{v_2}{2}-1\right)} e^{-(v_1 y_1 + v_2 y_2)/2}$$

Then the distribution function of the quotient (2.1.11) is

$$F_{v_1, v_2}(x) = \frac{v_1^{\frac{v_1}{2}} v_2^{\frac{v_2}{2}}}{2^{\left(\frac{v_1 + v_2}{2}\right)} \Gamma\left(\frac{v_1}{2}\right) \Gamma\left(\frac{v_2}{2}\right)} \int\!\!\int y_1^{\left(\frac{v_1}{2}-1\right)} y_2^{\left(\frac{v_2}{2}-1\right)} e^{-\left(\frac{v_1 y_1 + v_2 y_2}{2}\right)} dy_1\, dy_2$$

The region of integration is $0 \leq \dfrac{y_1}{y_2} \leq x$, $y_2 \geq 0$. By introducing the substitution

$$y_1 = tv \qquad y_2 = v$$

and by calculation of the Jacoby determinant

$$I = \begin{vmatrix} \dfrac{\partial y_1}{\partial t} & \dfrac{\partial y_1}{\partial v} \\[2ex] \dfrac{\partial y_2}{\partial t} & \dfrac{\partial y_2}{\partial v} \end{vmatrix} = v$$

we obtain the distribution function

$$
F_{v_1, v_2}(x) = \frac{\left(\frac{v_1}{v_2}\right)^{\left(\frac{v_1}{v_2}\right)} \Gamma\left(\frac{v_1 + v_2}{2}\right)}{\Gamma\left(\frac{v_1}{2}\right)\Gamma\left(\frac{v_2}{2}\right)} \int\limits_0^x \frac{t^{\left(\frac{v_1}{2}-1\right)}}{\left(1 + \frac{v_1}{v_2}t\right)^{\left(\frac{v_1 + v_2}{2}\right)}} \, dt
$$

The quotient (2.1.11) thus shows the Fisher-Snedecor distribution (F distribution in Appendix A):

$$
\frac{\delta_{\mathscr{g}}^2 + m\delta_{\mathscr{E}}^2}{\delta_{\mathscr{g}}^2} = F_{0, v_1, v_2}
$$

If $\delta_{\mathscr{E}}^2 \to 0$, then

$$
F_{0, v_1, v_2} \lessgtr F_{\alpha, v_1, v_2}
$$

and hypothesis H_0 will be accepted. Similarly, if

$$
F_{0, v_1, v_2} > F_{\alpha, v_1, v_2}
$$

hypothesis H_0 will be rejected.

In the actual case of the crystal structure set containing s structures for a certain central entity type, we obtain n sets of frequencies of interatomic vector lengths grouped in an unequal number of classes, i.e. a *non — orthogonal case*. In fact for a certain central entity type the matrix of frequencies (2.1.1) will have the form

$$
\begin{bmatrix}
x_{11} & x_{21} & x_{31} & \cdots & x_{n1} \\
x_{12} & x_{22} & x_{32} & \cdots & x_{n2} \\
x_{13} & x_{23} & x_{33} & \cdots & x_{n3} \\
\vdots & \vdots & \vdots & & \vdots \\
x_{1m_1} & x_{2m_2} & x_{3m_3} & \cdots & x_{nm_n}
\end{bmatrix}
$$

The formulae (2.1.9) and (2.1.10) for the calculation of point estimates of variances $\sigma_{\mathscr{g}}^2$ and $\sigma_{\mathscr{E}}^2$ will then obtain the form

$$
\delta_{\mathscr{g}}^2 = \sum_{i=1}^n m_i(x_{i.} - x_{..})^2/(n-1)
$$

$$
\delta_{\mathscr{E}}^2 = \sum_{i=1}^n \sum_{j=1}^{m_i} (x_{ij} - x_{i.})^2/k \sum_{i=1}^n (m_i - n) - \sum_{i=1}^n m_i(x_{i.} - x_{..})/k(n-1)
$$

where

$$x_{i.} = \frac{1}{m_i} \sum_{j=1}^{m_i} x_{ij} \qquad x_{..} = \frac{1}{N} \sum_{i=1}^{n} \sum_{j=1}^{m_i} x_{ij} \qquad N = \sum_{i=1}^{n} m_i$$

$$k = \frac{1}{n-1} \left(N - \frac{1}{N} \sum_{i=1}^{n} m_i^2 \right)$$

Interclass correlation coefficient is

$$\varrho = \frac{\mathrm{Cov}(x_{ij}, x_{ij})}{\mathrm{Var}(x_{ij})\,\mathrm{Var}(x_{ij})}$$

According to Eqs. (2.1.6) and (2.1.7) for this coefficient we have the relation

$$\varrho = \frac{\sigma_{\mathscr{E}}^2}{\sigma_{\mathscr{g}}^2 + \sigma_{\mathscr{E}}^2}$$

and its point estimate makes

$$v = \frac{\delta_{\mathscr{E}}^2}{\delta_{\mathscr{g}}^2 + \delta_{\mathscr{E}}^2} \tag{2.1.12}$$

In practical applications of variance analysis, if the estimate δ^2 comes to negative value, then v is put zero.

Table 2.1.1 presents the results of variance analysis of frequencies using program VARAN, which was made up on the basis of the described algorithm. The critical F values were computed applying subprogram QUAFIS [1]. The matrices of frequencies of the dimensions $(n \times m_n)$ were computed for unique sets of central monatomic entities using the DIRDOM program. The computations were done with computers MINSK 4030.1 (VARAN) and EC 1033 (DIRDOM).

Table 2.1.1. The results of variance analysis of frequencies of interatomic vector lengths for coordination compounds. The significance level is $\alpha = 0.05$

Central monatomic entity	n	ϑ [%]	$F_{0, v_1, v_2} \leqq F_{\alpha, v_1, v_2}$ 1 — yes 0 — no
Sc^{3+}	29	1.73	0
Ti^{2+}	44	1.30	0
Ti^{3+}	68	0.0	1
Ti^{4+}	196	0.0	1
V	14	4.69	0
V^{1+}	15	6.23	0
V^{2+}	18	17.45	0
V^{3+}	24	1.52	0
V^{4+}	60	0.44	0

Table 2.1.1 (continued)

Central monatomic entity	n	$\vartheta\,[\%]$	$F_{0,\,v_1,\,v_2} \leqq F_{\alpha,\,v_1,\,v_2}$ 1 — yes 0 — no
V^{5+}	63	0.91	0
Cr	190	0.10	0
Cr^{1+}	51	0.69	0
Cr^{2+}	125	0.0	1
Cr^{3+}	175	0.0	1
Mn^{1-}	37	1.32	0
Mn	141	0.0	1
Mn^{1+}	194	0.0	1
Mn^{2+}	563	0.0	1
Mn^{3+}	50	0.58	0
Fe^{2-}	68	2.86	0
Fe	631	0.0	1
Fe^{1+}	266	0.0	1
Fe^{2+}	613	0.0	1
Fe^{3+}	332	0.50	0
Co^{1-}	51	0.37	0
Co	296	0.0	1
Co^{1+}	163	0.0	1
Co^{2+}	596	0.0	1
Co^{3+}	732	0.0	1
Ni	122	0.74	0
Ni^{1+}	58	0.0	1
Ni^{2+}	1125	0.0	1
Ni^{3+}	31	1.78	0
Ni^{4+}	37	0.36	0
Cu^{1+}	460	0.0	1
Cu^{2+}	2174	0.0	1
Zn^{2+}	294	0.0	1
Y^{3+}	19	5.99	0
Zr^{2+}	20	2.34	0
Zr^{4+}	62	1.55	0
Nb^{1+}	11	6.37	0
Nb^{2+}	10	6.18	0
Nb^{3+}	13	4.07	0
Nb^{5+}	95	0.0	1
Mo	110	0.12	0
Mo^{1+}	100	0.0	1
Mo^{2+}	319	0.0	1
Mo^{3+}	94	0.0	1
Mo^{4+}	160	0.0	1
Mo^{5+}	184	0.0	1
Mo^{6+}	281	0.0	1
Tc^{3+}	10	4.70	0
Tc^{5+}	12	2.19	0
Ru	363	0.0	1
Ru^{1+}	21	17.94	0
Ru^{2+}	143	0.0	1
Ru^{3+}	26	1.36	0
Rh	299	0.0	1
Rh^{1+}	188	0.0	1
Rh^{2+}	56	0.61	0

Central monatomic entity	n	$\vartheta\,[\%]$	$F_{0,\,v_1,\,v_2} \leqq F_{\alpha,\,v_1,\,v_2}$ 1 — yes 0 — no
Rh^{3+}	117	8.93	0
Pd	61	2.97	0
Pd^{1+}	444	0.0	1
Pd^{2+}	454	0.0	1
Ag^{1+}	908	0.0	1
Cd^{2+}	824	0.0	1
La^{3+}	79	0.36	0
Ce^{3+}	60	0.0	1
Ce^{4+}	111	0.0	1
Pr^{3+}	35	1.99	0
Nd^{3+}	163	0.24	1
Sm^{3+}	11	4.34	0
Eu^{3+}	37	1.52	0
Gd^{3+}	58	1.00	0
Er^{3+}	121	0.0	1
Yb^{3+}	28	1.80	0
Lu^{3+}	10	4.26	0
Hf^{4+}	13	2.21	0
Ta^{3+}	31	1.23	0
Ta^{5+}	39	0.0	1
W	94	0.41	0
W^{1+}	22	2.33	0
W^{2+}	70	0.51	0
W^{3+}	48	0.58	0
W^{4+}	67	0.42	0
W^{5+}	16	4.32	0
W^{6+}	84	0.43	0
Re	17	4.94	0
Re^{1+}	66	0.13	0
Re^{2+}	16	3.86	0
Re^{3+}	96	0.0	1
Re^{4+}	12	5.97	0
Re^{5+}	34	0.72	0
Os	459	0.0	1
Os^{2+}	38	3.40	0
Os^{6+}	11	2.08	0
Ir	67	0.30	0
Ir^{1+}	67	0.60	0
Ir^{2+}	11	4.07	0
Ir^{3+}	79	0.0	1
Pt	201	0.0	1
Pt^{2+}	528	0.0	1
Pt^{4+}	39	0.86	0
Au^{1+}	101	0.0	1
Au^{3+}	56	0.69	0
Hg^{1+}	94	0.43	0
Hg^{2+}	869	0.0	1

2.2 Nonrigid Behavior of Central Monatomic Entities

Based on the results of variance analysis of interatomic vector lengths as presented in Table 2.1.1, we will further on limit study only to those central monatomic entities having significance on the whole variance just component $\sigma_\mathscr{G}^2$ (2.1.6). Figure 2.2.1 shows point diagram m_1 vs. m_2 for these entities. Suppose a normal twodimensional distribution of these data, where the estimate of the correlation coefficient will be $r_1 = 0.77$, this being higher than its critical value $r_{1\,\text{crit}} = 0.30$ for $\alpha = 0.05$ and $\mathrm{m} = 46$ [14]. Thus the hypothesis of the mutual independence of parameters m_1 and m_2 can be rejected and again their linear dependence can be

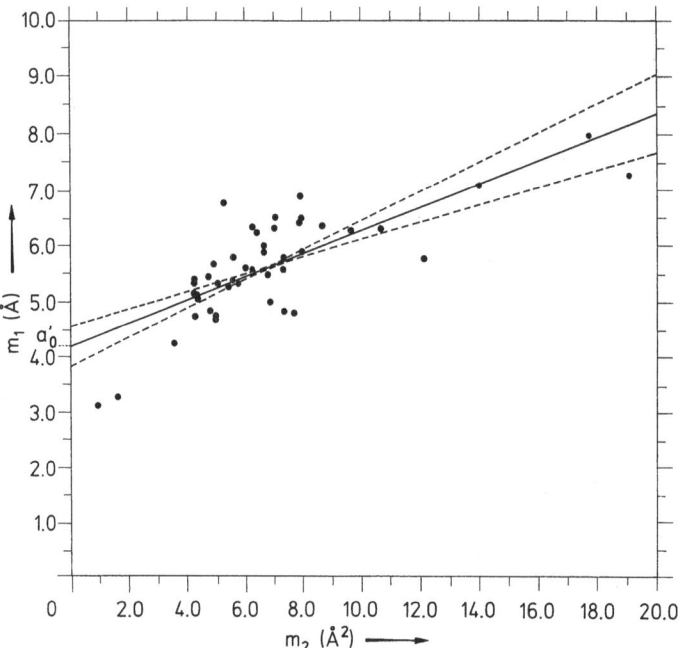

Fig. 2.2.1. Correlation of empirical estimates m_1 vs. m_2. Only the moments of interatomic vector lengths distributions showing significant averaging of extrinsic factors ($\sigma_\mathscr{E}^2 \to 0$) are included. The equation of regression line is $m_1 = 4.25 + (0.21)\,m_2$. The confidence intervals of linear regression are shown by dashed curves. Points within these confidence intervals belong to Ti^{2+}, Mn^{1+}, Fe, Fe^{2+}, Co^{3+}, Ni^{1+}, Cu^{1+}, Nb^{5+}, Mo^{2+}, Rh, Ag^{1+}, Ce^{3+}, Nd^{3+} and Re^{3+}.

assumed. The regression function for this case is $m_1 = a_0' + b'm_2$. By the least squares method we obtain $a_0' = 4.25$ Å and $b' = 0.21$ Å^{-1}. Though for us a_0' is at a minimum distance from the position of the central entity, from which the probability model Γ may be accepted, the point variance around the regression line in Fig. 2.2.1 indicates a statistical variance also of this value. Suppose that pairs (m_{2i}, m_{1i}) are the studied pairs of random variables (X, Y) with normal distribution, we may then introduce the testing quantity

$$T_{m-2} = \frac{Y - Y_0}{S_1}$$

where

$$S_1 = S' \sqrt{\frac{1}{m} + \frac{\bar{X}^2}{(m-1)\,S_x^2}}$$

$$Y = a_0' + b'X$$

$$S' = \frac{1}{m-2} \sum_{i=1}^{m} [m_{1i} - (a_0' + b'm_{1i})]^2$$

This variable has t distribution with $m - 2$ degrees of freedom[1]. Then one can formulate the hypothesis of

$$H_0 : Y = Y_0 \quad \text{against} \quad Y \neq Y_0$$

From the inequality

$$\Pr\left(-t_{m-2;\,1-\alpha/2} < T_{m-2} < t_{m-2;\,1-\alpha/2}\right) = 1 - \alpha$$

the confidence interval of variable Y can be determined:

$$a_0' + b'X - S_1 t_{n-2;\,1-\alpha/2} < Y < a_0' + b'X + S_1 t_{m-2;\,1-\alpha/2} \qquad (2.2.1)$$

The limits of intervals (2.2.1) are shown in Fig. 2.2.1 for $\alpha = 0.05$ by dashed lines above and below the regression line. The points lying outside this region exhibit a significant deflection from linear dependence. For these central entities the Γ probability model is less suitable. As Fig. 2.2.1 shows the minimum length a_0' also exhibits some variance. Its confidence interval reckoned by the inequality (2.2.1) and the inserting of $X = 0$, makes 3.84–4.66 Å.

The variance of interatomic vector lengths with a common origin in the position of a certain central monatomic entity M^{z+} for averaged extrinsic factors shall be called its *crystallographic nonrigidity*. The measure of this intrinsic behavior of the central entity is the estimate of the second central moment (m_2). Similarly as for variance, the third central moment of the transformed random variable $W = X - a_0'$ the identity $v_{3W} = v_{3X}$ can also be proved. For parameters of Γ distribution the relations hold (Table 1.1.1) $\gamma^2 = 4/\tau$ and $\sigma^2 = \tau/\lambda^2$, from which after elimination of τ we will obtain

$$\gamma^2 = \frac{4}{\sigma^2 \lambda^2} \qquad (2.2.2)$$

---

[1] For evidence see [15].

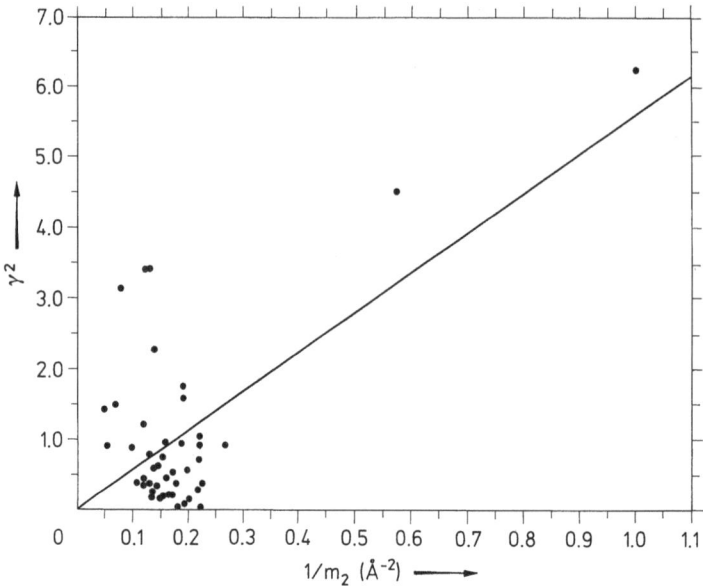

Fig. 2.2.2. Correlation γ^2 *vs.* $1/m_2$. Only the moments (m_2 and m_3) of interatomic vector lengths distributions showing significant averaging of extrinsic factors ($\sigma^2 \to 0$) are included. The equation of regression line is $\gamma^2 = 0.02 + (5.30)/m_2$

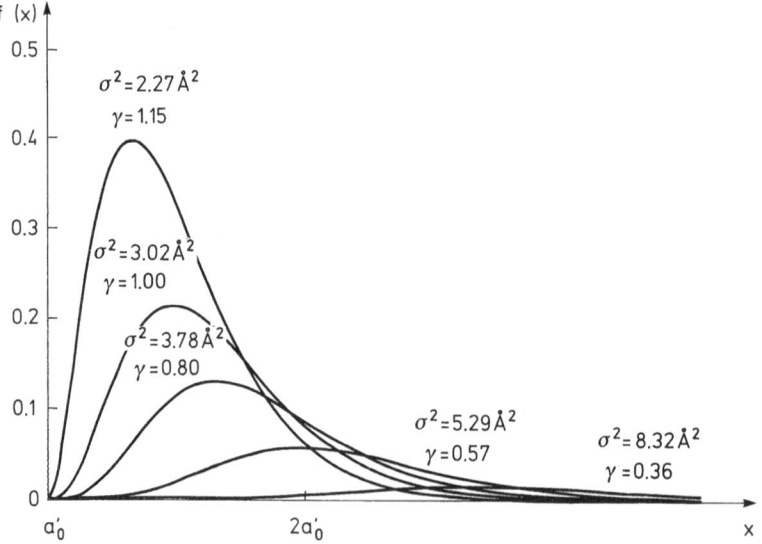

Fig. 2.2.3. Graphical outline of the dependence between the shape of density function of interatomic vector lengths and its variance (σ^2) or asymmetry (γ). Γ probability model was used; $\lambda = 1.15 \text{ Å}^{-1}$.

Figure 2.2.2 presents the point diagram of correlation γ^2 vs. $1/m_2$. The correlation coefficient is $r_3 = 0.65$ being again much higher than its critical value. The regression line has the equation of $\gamma^2 = a'' + b''/m_2$, where $a'' = 0.02$ and $b'' = 5.30 \, \text{Å}^2$. The confidence interval of parameter a'' for $\alpha = 0.05$ is $(-1.29, 1.34)$. Thus in agreement with the presupposed relation (2.2.2) it can be considered zero.

Accepting probability model Γ for radial distribution of interatomic vectors, the crystallographic nonrigidity of central monatomic entity (m_2) expresses its tendency to spread interatomic vector lengths about their first moment. Thus it expresses the tendency of central entity to violate the vector equilibrium in the sense of VEP. The empirical estimates of the first moments (m_1) are listed in Table 1.1.3. The smaller the asymmetry of interatomic vectors radial distribution of a certain central entity, the greater is its tendency to form the highest possible quantity of shortest contacts with the other entities of the structure. This relation is simulated in Fig. 2.2.3. Parameter λ estimated by the explained regression analysis is $\lambda = b''/2 \doteq 1.15 \, \text{Å}^{-1}$. The measure of this property of central entity (γ) thus

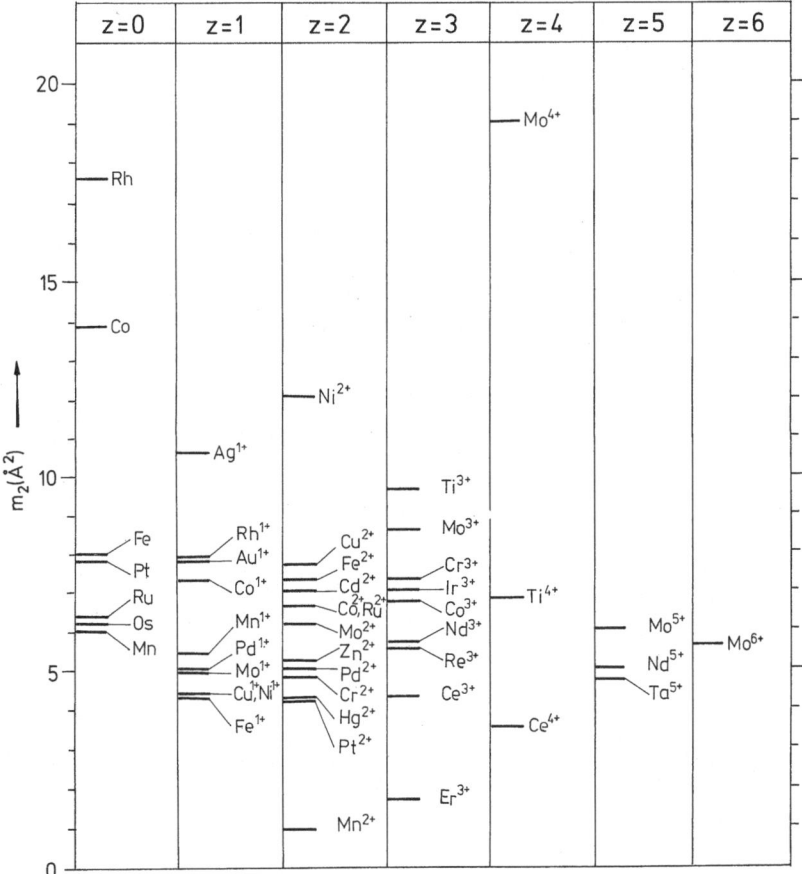

Fig. 2.2.4. Level diagram of crystallographic non-rigidity of the central monatomic entities.

expresses its tendency to turn off from the connection principle (Chap. 1.2). Level diagrams of m_2 and γ values are shown in Figs. 2.2.4 and 2.2.5.

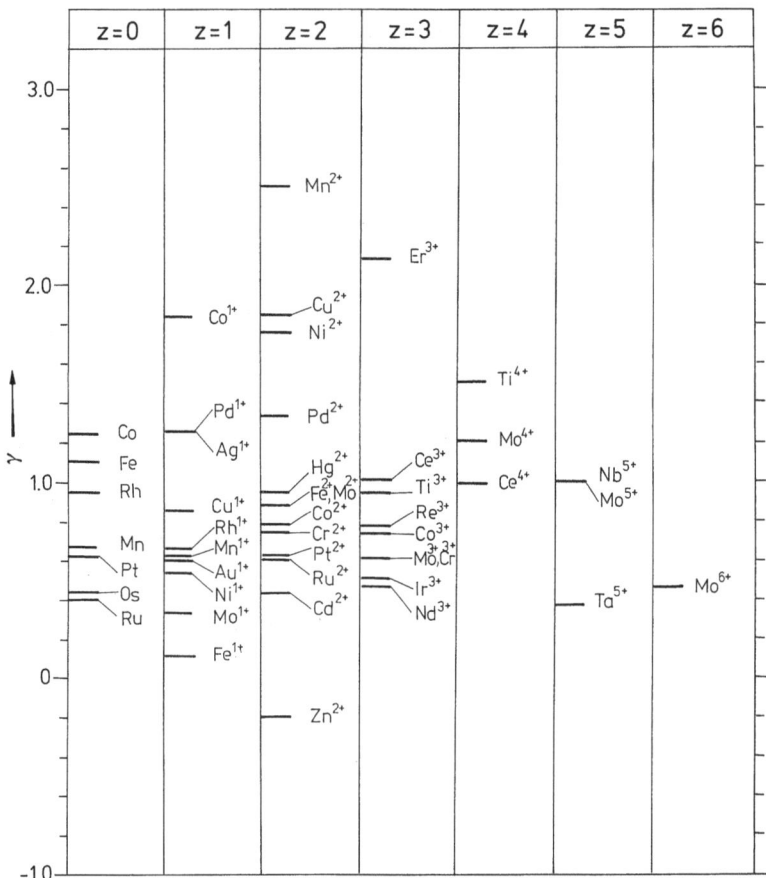

Fig. 2.2.5. Level diagram of values γ.

Glossary of Symbols

n	Sample size; number of interatomic vector lengths; number of central monatomic entities of the same type from crystallochemical units (unique set of residues in structure) of the crystal structures
S	Enthropy
p_i	Occurance probability of interatomic vector length in the ith class
$\boldsymbol{a}, \boldsymbol{b}, \boldsymbol{c}$	Unit cell vectors
$\boldsymbol{r}$	Radius vector
$\mathbb{H}$	Hamiltonian
Ψ	Wave function
E	Eigenvalue of Hamiltonian
U	Potential energy

V	Potential of crystal field
$\hbar$	Planck constant devided by 2π; $\hbar = h/2\pi$
Δ	Operator $\Delta = \dfrac{\partial^2}{\partial x^2} + \dfrac{\partial^2}{\partial y^2} + \dfrac{\partial^2}{\partial x^2}$
a^*, b^*, c^*	Unit cell vectors of reciprocal lattice
g^*	Reciprocal lattice vector
Ω	Unit cell volume of a crystal
$\varphi_i(000)$	Average value of crystal field potential
s	Number of the crystal structures
m	Number of data classes
x_{ij}	Frequence of interatomic vector lengths
$\mathscr{I}_{ij}$	Intrinsic contribution to x_{ij}
$\mathscr{E}_{ij}$	Extrinsic contribution to x_{ij}
p_{ij}	Probability of random variable x_{ij}
μ	General mean value of variables x_{ij}
$\delta_{\mathscr{I}}^2$	Point estimate of intrinsic variance
$\delta_{\mathscr{E}}^2$	Point estimate of extrinsic variance
$\Gamma(x)$	Γ function
F_{α, ν_1, ν_2}	Critical value of F distributed variable with significance level α and ν_1, ν_2 degrees of freedom
ϱ	Interclass correlation coefficient
υ	Point estimate of interclass correlation coefficient
r_1	Correlation coefficient of m_1 vs. m_2
a_0'	Minimum interatomic vector length for Γ probability model
b'	Coefficient of regression line m_1 vs. m_2
$\mathfrak{m}$	Number of points of correlaton m_1 vs. m_2 for the central entities having significant averaging of extrinsic factors
τ, λ	Parameters of Γ probability model
r_3	Correlation coefficient of γ^2 vs. $1/m_2$
a'', b''	Coefficients of regression line γ^2 vs. $1/m_2$

2.3 References

1. Program QUAFIS, SIEMENS PBS 4004, MEM METHODENBANK 1973
2. Shannon, C. E.: Bell Syst. Tech. J. **27**, 397 (1948)
3. Shannon, C. E.: Bell Syst. Tech. J. **27**, 623 (1948)
4. Jaynes, E. T.: Phys. Rev. **106**, 620 (1957)
5. Livesey, K. A., Skilling, J.: Acta Crystallogr., Sect. A, **41**, 113 (1985)
6. Scheffé, H.: The Analysis of Variance, New York, John Wiley & Sons 1958
7. Guenther, W. C.: Analysis of Variance, New York, Prentice-Hall Inc., Englewood Cliffs 1964
8. Kempthorne, O.: The Design and Analysis of Experiments, New York, Wiley & Sons 1952
9. Ahrens, H.: Varianzanalyse, Berlin, Akademie Verlag 1967
10. Dixon, W. I., Massey, F. J.: Introduction to Statistical Analysis, New York, McGraw-Hill Book Comp. 1969³
11. Henderson, C. R.: Biometrics **9**, 226 (1953)
12. Crump, S. L.: Biometrics **7**, 1 (1951)
13. Eisenhart, Ch.: Biometrics **3**, 1 (1947)
14. Conover, W. J.: Practical Nonparametric Statistics, New York, Wiley & Sons 1971
15. Heinhold, J., Gaede, K. W.: Ingenieur — Statistik, München, Oldenbourg 1968²

3 Statistics of Crystallographic Sites of Central Monatomic Entities

Nonrigid properties of monatomic entities manifest themselves in the whole geometry of their environs. Until now we studied the statistics of these geometries using radial distributions of interatomic vectors. The classification of surroundings of a certain central monatomic entity M^{z+} can be based also on their symmetry. It has group — theoretical properties and it can be described by means of non — crystallographic symmetry point groups, of which there are infinitely many. There is, however, a finite amount of crystallographic point groups, viz. 2 one-dimensional, 10 two-dimensional and 32 three-dimensional point groups. The latest of them also describe the site symmetry of monatomic entity in the crystal structure and thus also the symmetry of its surroundings. Surroundings of monatomic entity means here, however, not only the near surroundings usually consisting of coordinating ligands, but the complete surroundings containing all other non — coordinating entities of the structure. The symmetry of an ideal crystal structure — of discontinuum, is described by one of 230 space groups. The site symmetry of each monatomic entity of the structure thus has a certain connection with the relative positions of entities which may be classified based on the conception of *lattice complexes.*

The equidistance of interatomic vectors with common origin in point position of the central monatomic entity is also dependent on its site symmetry. We will call the equivalence of interatomic vector lengths expressed by the symmetry of site from the central monatomic entity the *crystallographic equidistance.* In the sense of *VEP* and Laves' principle of maximum symmetry, each structure entity should exhibit the tendency to occupy sites with the greatest symmetry. For coordination compounds, however, a significant of nuclei of monatomic entities forming the surroundings of the central entity is given in the existence of degenerate electron states. Left to themselves such states are not stable and usually their lifting takes place leading to a lowering of symmetry from surroundings of the central entity. A sufficiently great amount of data on the sites of central monatomic entities of coordination compounds stored in *Cambridge Crystallographic Data Files* affords the presuppositions to compute the statistics of occupation of certain crystallographic sites by monatomic entities. The classification of these sites, with respect to the degeneration of their irreducible representations, can provide information on the significance of lifting the degeneration of electronic states for the non — rigid properties of central monatomic entities in coordination compounds. From these points of view we will disregrad the orientation of symmetry elements of point groups at classification of crystallographic sites with respect to the crystallographic coordinate system.

3.1 Crystallographic Sites

Each point of crystal lattice called node has its own symmetry. The symmetry operation of a space group, mapping the lattice node A onto itself, is the point symmetry operation. These operations form the 32 crystallographic point groups tabulated in *International Tables for Crystallography* [1, 2]. Thus, one may say that the group of all point group operations of a space group, mapping point A onto itself, is the site-symmetry group of the space group under consideration. For example, space group P2/m contains symmetry centers in the positions

0,0,0	1/2,1/2,0	0,1/2,1/2	1/2,0,1/2
1/2,1/2,1/2	0,0,1/2	1/2,0,0	0,1/2,0

None of these symmetry centers are, however, a site-symmetry group of space group P2/m, since they are not groups consisting of all symmetry operations of the space group considered, which map point A onto itself. The site-symmetry groups of space group P2/m in the above-mentioned positions are $2/m-C_{2h}$ [1]. The set of N equivalent points are in at least one of 32 crystallographic point groups which can occupy positions in the discontinuum and is therefore called *crystallographic site set* or *N-pointer* [3]. Each of its site is invariant against some symmetry operations of point group, forming its subgroup, called the *site symmetry*. Thus *crystallographic site* (site) has its position on some symmetry element, i.e. a special position, or the general position with symmetry $1-C_1$. Only one site of discontinuum can show symmetry $\bar{1}-C_i$. This point is called according to [3], the *highest symmetry point*. The intersection of all elements (S) of site symmetry can be a point, line, plane or space. The number of independent positional parameters f_S of points contained in S is the *number of degrees of freedom*. Apparently it is number 0, 1, 2 or 3, which determines whether set S is a point, line, plane or space. Site symmetries of all space groups are listed in [3] and [4].

  The points of each N-pointer are placed on the surface of a sphere, with its center in the origin of the coordinate system being identical with the highest symmetry point. If we consider these points for the poles of planes, then site symmetry can be described by means of polyhedra which are tabulated in stereographical projections in [3]. If we displace, the point placed in the fixed origin from its original position, then symmetry operations will generate N_1 equivalent points (N_1-pointer). We say that the point in the origin is *split* to N_1 points. After the deviation of the point out of the origin, it can be found on some symmetry elements of the point group. This is then the case of a *partial splitting*. By further displacement of the point from this symmetry element, we generally obtain site symmetry $1-C_1$. We say that we achieved *complete splitting*. Table 3.1.1 presents point splittings for all site symmetries.

  Crystallographic sites have certain positions with respect to their crystallographic origin. The number of equivalent sites m_S in one unit cell is the *multiplicity of this site*. Then the question arises, under which conditions is a site symmetry S_1 contained in another site symmetry S of the same space group. This means that symmetry group S_1 should be a subgroup of S. Thus e.g. discontinuum of the

---

[1] Herman — Maugin (left) and Schoenflies symbol (right).

Table 3.1.1. Points splitting of site-symmetries [3] classified according to the site-symmetry after the first, second and third splitting

Point group	1st splitting	2nd splitting	3rd splitting
$1\text{-}C_1$ $\bar{1}\text{-}C_i$ $\bar{4}\text{-}S_4$ $\bar{3}\text{-}C_{3i}$	$1\text{-}C_1$		
$m\text{-}C_s$ $2/m\text{-}C_{2h}$ $\bar{6}\text{-}C_{3h}$ $4/m\text{-}C_{4h}$ $\bar{4}2m\text{-}D_{2d}$ $6/m\text{-}C_{6n}$ $\bar{3}m\text{-}D_{3d}$	$m\text{-}C_s$	$1\text{-}C_1$	
$2\text{-}C_2$ $222\text{-}D_2$ $32\text{-}D_3$ $422\text{-}D_4$ $622\text{-}D_6$ $23\text{-}T$	$2\text{-}C_2$	$1\text{-}C_1$	
$3\text{-}C_3$	$3\text{-}C_3$	$1\text{-}C_1$	
$4\text{-}C_4$ $432\text{-}O$	$4\text{-}C_4$	$1\text{-}C_1$	
$6\text{-}C_6$ $mm2\text{-}C_{2v}$ $3m\text{-}C_{3v}$ $mmm\text{-}D_{2h}$ $\bar{6}m2\text{-}D_{3h}$ $4/mmm\text{-}D_{4h}$ $6/mmm\text{-}D_{6h}$ $m\bar{3}\text{-}T_h$ $\bar{4}3m\text{-}T_d$	$6\text{-}C_6$ $mm2\text{-}C_{2v}$	$1\text{-}C_1$ $m\text{-}C_s$	 $1\text{-}C_1$
$4mm\text{-}C_{4v}$ $m\bar{3}m\text{-}O_h$	$4mm\text{-}C_{4v}$	$m\text{-}C_s$	$1\text{-}C_1$
$6mm\text{-}C_{6v}$	$6mm\text{-}C_{6v}$	$m\text{-}C_s$	$1\text{-}C_1$

space group P2/m covers the symmetry sites $1\text{-}C_1$, $2\text{-}C_2$ and $m\text{-}C_m$, being subgroups of the site symmetry $2/m\text{-}C_{2h}$. The multiplicity of these sites are 4, 2, 2 and 1. When group S_1 is contained in group S, then the relation holds [5]

$$f_{\mathbf{S}} \leqq f_{\mathbf{S}_1} - 1 \qquad\qquad m_{\mathbf{S}} \leqq m_{\mathbf{S}_1}/2 \qquad\qquad (3.1.1)$$

Figure 3.1.1 shows the relations between point groups and their subgroups. According to the relations (3.1.1) this Figure follows as consequence:

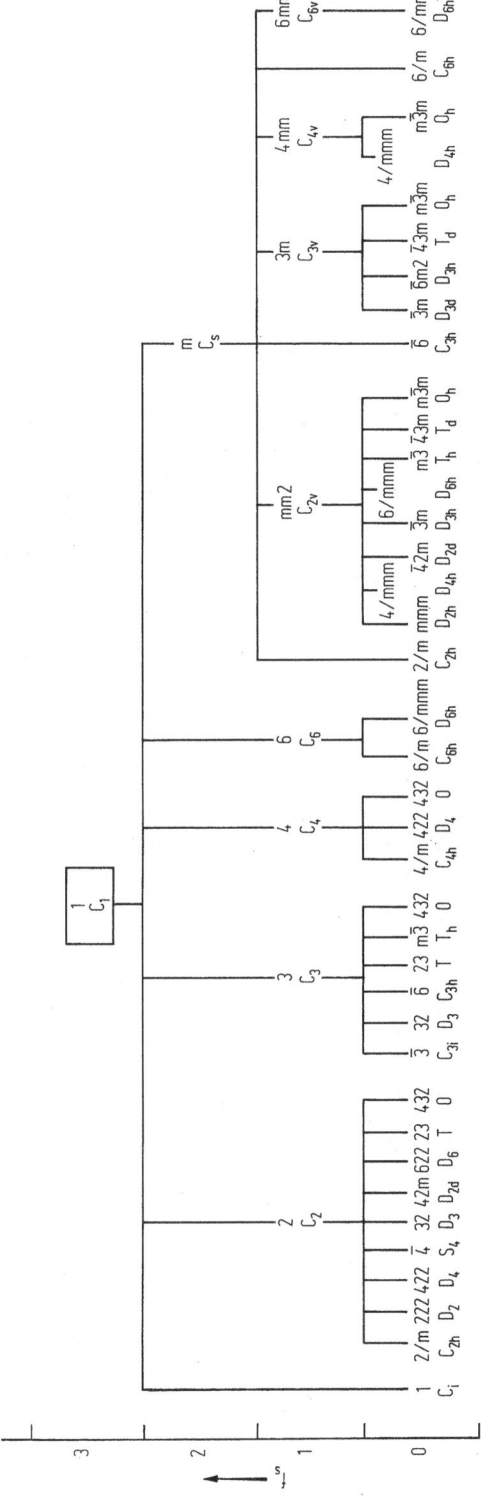

Fig. 3.1.1. Relations between point groups. Subgroups are ordered by their degrees of freedom (f).

Table 3.1.2. Main point groups of the space groups. In parantheses are given multicity (m_S) and degree of freedom (f_S) of site. Main groups of each space group are arranged in alphabetic order of their Wickoff letters in *International Tables for Crystallography* [2], Part II, pp. 102–707

Space group symbol	No.	Main groups
P1-C_1^1	1	1-C_1 (1,3)
P$\bar{1}$-C_i^1	2	$\bar{1}$-C_i (1,0)
P2-C_2^1	3	2-C_2 (1,1)
P2_1-C_2^2	4	1-C_1 (2,3)
C2-C_2^3	5	2-C_2 (2,1)
Pm-C_s^1	6	m-C_s (1,2)
Pc-C_s^2	7	1-C_1 (2,3)
Cm-C_s^3	8	m-C_s (2,2)
Cc-C_s^4	9	1-C_1 (4,3)
P2/m-C_{2h}^1	10	2/m-C_{2h} (1,0)
P2_1/m-C_{2h}^2	11	$\bar{1}$-C_i (2,0); m-C_s (2,2)
C2/m-C_{2h}^3	12	2/m-C_{2h} (2,0)
P2/c-C_{2h}^4	13	$\bar{1}$-C_i (2,0); 2-C_2 (2,1)
P2_1/c-C_{2h}^5	14	$\bar{1}$-C_i (2,0)
C2/c-C_{2h}^6	15	$\bar{1}$-C_i (4,0); 2-C_2 (4,1)
P222-D_2^1	16	222-D_2 (1,0)
P222_1-D_2^2	17	2-C_2 (2,1)
P$2_1 2_1 2$-D_2^3	18	2-C_2 (2,1)
P$2_1 2_1 2_1$-D_2^4	19	1-C_1 (4,3)
C222_1-D_2^5	20	2-C_2 (4,1)
C222-D_2^6	21	222-D_2 (2,0)
F222-D_2^7	22	222-D_2 (4,0)
I222-D_2^8	23	222-D_2 (2,0)
I$2_1 2_1 2_1$-D_2^9	24	2-C_2 (4,1)
Pmm2-C_{2v}^1	25	mm2-C_{2v} (1,1)
Pmc2_1-C_{2v}^2	26	m-C_s (2,2)
Pcc2-C_{2v}^3	27	2-C_2 (2,1)
Pma2-C_{2v}^4	28	2-C_2 (2,1); m-C_s (2,2)
Pca2_1-C_{2v}^5	29	1-C_1 (4,3)
Pnc2-C_{2v}^6	30	2-C_2 (2,1)
Pmn2_1-C_{2v}^7	31	m-C_s (2,2)
Pba2-C_{2v}^8	32	2-C_2 (2,1)
Pna2_1-C_{2v}^9	33	1-C_1 (4,3)
Pnn2-C_{2v}^{10}	34	2-C_2 (2,1)
Cmm2-C_{2v}^{11}	35	mm2-C_{2v} (2,1)
Cmc2_1-C_{2v}^{12}	36	m-C_s (4,2)
Ccc2-C_{2v}^{13}	37	2-C_2 (4,1)
Amm2-C_{2v}^{14}	38	mm2-C_{2v} (2,1)
Abm2-C_{2v}^{15}	39	2-C_2 (4,1); m-C_s (4,2)
Ama2-C_{2v}^{16}	40	2-C_2 (4,1); m-C_s (4,2)
Aba2-C_{2v}^{17}	41	2-C_2 (4,1)

Space group symbol	No.	Main groups
Fmm2-C_{2v}^{18}	42	mm2-C_{2v} (4,1)
Fdd2-C_{2v}^{19}	43	2-C_2 (8,1)
Imm2-C_{2v}^{20}	44	mm2-C_{2v} (2,1)
Iba2-C_{2v}^{21}	45	2-C_2 (4,1)
Ima2-C_{2v}^{22}	46	2-C_s (4,1); m-C_2 (4,2)
Pmmm-D_{2h}^{1}	47	mmm-D_{2h} (1,0)
Pnnn-D_{2h}^{2}	48	222-D_2 (2,0); $\bar{1}$-C_i (4,0)
Pccm-D_{2h}^{3}	49	2/m-C_{2h} (2,0); 222-D_2 (2,0)
Pban-D_{2h}^{4}	50	222-D_2 (2,0); $\bar{1}$-C_i (4,0)
Pmma-D_{2h}^{5}	51	2/m-C_{2h} (2,0); mm2-C_{2v} (2,1)
Pnna-D_{2h}^{6}	52	$\bar{1}$-C_i (4,0); 2-C_2 (4,1)
Pmna-D_{2h}^{7}	53	2/m-C_{2h} (2,0)
Pcca-D_{2h}^{8}	54	$\bar{1}$-C_i (4,0); 2-C_2 (4,1)
Pbam-D_{2h}^{9}	55	2/m-C_{2h} (2,0)
Pccn-D_{2h}^{10}	56	$\bar{1}$-C_i (4,0); 2-C_2 (4,1)
Pbcm-D_{2h}^{11}	57	$\bar{1}$-C_i (4,0); 2-C_2 (4,1); m-C_s (4,2)
Pnnm-D_{2h}^{12}	58	2/m-C_2 (2,0)
Pmmn-D_{2h}^{13}	59	mm2-C_{2v} (2,1); $\bar{1}$-C_i (4,0)
Pbcn-D_{2h}^{14}	60	$\bar{1}$-C_i (4,0); 2-C_2 (4,1)
Pbca-D_{2h}^{15}	61	$\bar{1}$-C_i (4,0)
Pnma-D_{2h}^{16}	62	$\bar{1}$-C_i (4,0); m-C_s (4,2)
Cmcm-D_{2h}^{17}	63	2/m-C_{2h} (4,0); mm2-C_{2v} (4,1)
Cmca-D_{2h}^{18}	64	2/m-C_{2h} (4,0)
Cmmm-D_{2h}^{19}	65	mmm-D_{2h} (2,0)
Cccm-D_{2h}^{20}	66	222-D_2 (4,0); 2/m-C_{2h} (4,0)
Cmma-D_{2h}^{21}	67	222-D_2 (4,0); 2/m-C_{2h} (4,0); mm2-C_{2v} (4,1)
Ccca-D_{2h}^{22}	68	222-D_2 (4,0); $\bar{1}$-C_i (8,0)
Fmmm-D_{2h}^{23}	69	mmm-D_{2h} (4,0)
Fddd-D_{2h}^{24}	70	222-D_2 (8,0); $\bar{1}$-C_i (16,0)
Immm-D_{2h}^{25}	71	mmm-D_{2h} (2,0)
Ibam-D_{2h}^{26}	72	222-D_2 (4,0); 2/m-C_{2h} (4,0)
Ibca-D_{2h}^{27}	73	$\bar{1}$-C_i (8,0); 2-C_2 (8,1)
Imma-D_{2h}^{28}	74	2/m-C_h (4,0); mm2-C_{2v} (4,1)
P$_4$-C_4^{1}	75	4-C_4 (1,1)
P4$_1$-C_4^{2}	76	1-C_1 (4,3)
P4$_2$-C_4^{3}	77	2-C_2 (2,1)
P4$_3$-C_4^{4}	78	1-C_1 (4,3)
I4-C_4^{5}	79	4-C_4 (2,1)
I4$_1$-C_4^{6}	80	2-C_2 (4,1)
P$\bar{4}$-S_4^{1}	81	$\bar{4}$-S_4 (1,0)
I$\bar{4}$-S_4^{2}	82	$\bar{4}$-S_4 (2,0)
P4/m-C_{4h}^{1}	83	4/m-C_{4h} (1,0)
P4$_2$/m-C_{4h}^{2}	84	2/m-C_{2h} (2,0); $\bar{4}$-S_4 (2,0)
P4/n-C_{4h}^{3}	85	$\bar{4}$-S_4 (2,0); 4-C_4 (2,1); $\bar{1}$-C_i (4,0)

Table 3.1.2 (continued)

Space group symbol	No.	Main groups
$P4_2/n\text{-}C_{4h}^4$	86	$\bar{4}\text{-}S_4\ (2,0);\ \bar{1}\text{-}C_i\ (4,0)$
$I4/m\text{-}C_{4h}^5$	87	$4/m\text{-}C_{4h}\ (2,0)$
$I4_1/a\text{-}C_{4h}^6$	88	$\bar{4}\text{-}S_4\ (4,0);\ \bar{1}\text{-}C_i\ (8,0)$
$P422\text{-}D_4^1$	89	$422\text{-}D_4\ (1,0)$
$P42_12\text{-}D_4^2$	90	$222\text{-}D_2\ (2,0);\ 4\text{-}C_4\ (2,1)$
$P4_122\text{-}D_4^3$	91	$2\text{-}C_2\ (4,1)$
$P4_12_12\text{-}D_4^4$	92	$2\text{-}C_2\ (4,1)$
$P4_222\text{-}D_4^5$	93	$222\text{-}D_2\ (2,0)$
$P4_22_12\text{-}D_4^6$	94	$222\text{-}D_2\ (2,0)$
$P4_322\text{-}D_4^7$	95	$2\text{-}C_2\ (4,1)$
$P4_32_12\text{-}D_4^8$	96	$2\text{-}C_2\ (4,1)$
$I422\text{-}D_4^9$	97	$422\text{-}D_4\ (2,0)$
$I4_122\text{-}D_4^{10}$	98	$222\text{-}D_2\ (4,0)$
$P4mm\text{-}C_{4v}^1$	99	$4mm\text{-}C_{4v}\ (1,1);\ mm2\text{-}C_{2v}\ (2,1)$
$P4bm\text{-}C_{4v}^2$	100	$4\text{-}C_4\ (2,1);\ mm2\text{-}C_{2v}\ (2,1)$
$P4_2cm\text{-}C_{4v}^3$	101	$mm2\text{-}C_{2v}\ (2,1)$
$P4_2nm\text{-}C_{4v}^4$	102	$mm2\text{-}C_{2v}\ (2,1)$
$P4cc\text{-}C_{4v}^5$	103	$4\text{-}C_4\ (2,1)$
$P4nc\text{-}C_{4v}^6$	104	$4\text{-}C_4\ (2,1)$
$P4_2mc\text{-}C_{4v}^7$	105	$mm2\text{-}C_{2v}\ (2,1)$
$P4_2bc\text{-}C_{4v}^8$	106	$2\text{-}C_2\ (4,1)$
$I4mm\text{-}C_{4v}^9$	107	$4mm\text{-}C_{4v}\ (2,1);\ mm2\text{-}C_{2v}\ (4,1)$
$I4cm\text{-}C_{4v}^{10}$	108	$4\text{-}C_4\ (4,1);\ mm2\text{-}C_{2v}\ (4,1)$
$I4_1md\text{-}C_{4v}^{11}$	109	$mm2\text{-}C_{2v}\ (4,1)$
$I4_1cd\text{-}C_{4v}^{12}$	110	$2\text{-}C_2\ (8,1)$
$P\bar{4}2m\text{-}D_{2d}^1$	111	$\bar{4}2m\text{-}D_{2d}\ (1,0)$
$P\bar{4}2c\text{-}D_{2d}^2$	112	$222\text{-}D_2\ (2,0);\ \bar{4}\text{-}S_4\ (2,0)$
$P\bar{4}2_1m\text{-}D_{2d}^3$	113	$\bar{4}\text{-}S_4\ (2,0);\ mm2\text{-}C_{4v}\ (2,1)$
$P\bar{4}2_1c\text{-}D_{2d}^4$	114	$\bar{4}\text{-}S_4\ (2,0)$
$P\bar{4}m2\text{-}D_{2d}^5$	115	$\bar{4}2m\text{-}D_{2d}\ (1,0)$
$P\bar{4}c2\text{-}D_{2d}^6$	116	$222\text{-}D_2\ (2,0);\ \bar{4}\text{-}S_4\ (2,0)$
$P\bar{4}b2\text{-}D_{2d}^7$	117	$\bar{4}\text{-}S_4\ (2,0);\ 222\text{-}D_2\ (2,0)$
$P\bar{4}n2\text{-}D_{2d}^8$	118	$\bar{4}\text{-}S_4\ (2,0);\ 222\text{-}D_2\ (2,0)$
$I\bar{4}m2\text{-}D_{2d}^9$	119	$\bar{4}2m\text{-}D_{2d}\ (2,0)$
$I\bar{4}c2\text{-}D_{2d}^{10}$	120	$222\text{-}D_2\ (4,0);\ \bar{4}\text{-}S_4\ (4,0)$
$I\bar{4}2m\text{-}D_{2d}^{11}$	121	$\bar{4}2m\text{-}D_{2d}\ (2,0)$
$I\bar{4}2d\text{-}D_{2d}^{12}$	122	$\bar{4}\text{-}S_4\ (4,0)$
$P4/mmm\text{-}D_{4h}^1$	123	$4/mmm\text{-}D_{4h}\ (1,0)$
$P4/mcc\text{-}D_{4h}^2$	124	$422\text{-}D_4\ (2,0);\ 4/m\text{-}C_{4h}\ (2,0)$
$P4/nbm\text{-}D_{4h}^3$	125	$422\text{-}D_4\ (2,0);\ \bar{4}2m\text{-}D_{2d}\ (2,0);\ 2/m\text{-}C_{2h}\ (4,0)$
$P4/nnc\text{-}D_{4h}^4$	126	$422\text{-}D_4\ (2,0);\ \bar{4}\text{-}S_4\ (4,0);\ \bar{1}\text{-}C_i\ (8,0)$
$P4/mbm\text{-}D_{4h}^5$	127	$4/m\text{-}C_{4h}\ (2,0);\ mmm\text{-}D_{2h}\ (2,0)$
$P4mnc\text{-}D_{4h}^6$	128	$4/m\text{-}C_{4h}\ (2,0);\ 222\text{-}D_2\ (4,0)$

Space group symbol	No.	Main groups
P4/nmm-D_{4h}^7	129	$\bar{4}2m$-D_{2d} (2,0); 4mm-C_{4v} (2,1); 2/m-C_{2v} (4,0)
P4/ncc-D_{4h}^8	130	222-D_2 (4,0); $\bar{4}$-S_4 (4,0); 4-C_4 (4,0); $\bar{1}$-C_i (8,0)
P4$_2$/mmc-D_{4h}^9	131	mmm-D_{2h} (2,0); $\bar{4}2m$-D_{2d} (2,0)
P4$_2$/mcm-D_{4h}^{10}	132	mmm-D_{2h} (2,0); $\bar{4}2m$-D_{2d} (2,0); 2/m-C_{2h} (4,0)
P4$_2$/nbc-D_{4h}^{11}	133	222-D_2 (4,0); $\bar{4}$-S_4 (4,0); $\bar{1}$-C_i (8,0)
P4$_2$/nnm-D_{4h}^{12}	134	$\bar{4}2m$-D_{2d} (2,0); 222-D_2 (4,0); 2/m-C_{2h} (4,0)
P4$_2$/mbc-D_{4h}^{13}	135	2/m-C_{2h} (4,0); $\bar{4}$-S_4 (4,0); 222-D_2 (4,0)
P4$_2$/mnm-D_{4h}^{14}	136	mmm-D_{2h} (2,0); $\bar{4}$-S_4 (4,0)
P4$_2$/nmc-D_{4h}^{15}	137	$\bar{4}2m$-D_{2d} (2,0); $\bar{1}$-C_i (8,0)
P4$_2$/ncm-D_{4h}^{16}	138	222-D_2 (4,0); $\bar{4}$-S_4 (4,0); 2/m-C_{2h} (4,0); mm2-C_{2v} (4,1)
I4/mmm-D_{4h}^{17}	139	4/mmm-D_{4h} (2,0)
I4/mcm-D_{4h}^{18}	140	422-D_4 (4,0); $\bar{4}2m$-D_{2d} (4,0); 4/m-C_{4h} (4,0); mmm-D_{2h} (4,0)
I4$_1$/amd-D_{4h}^{19}	141	$\bar{4}2m$-D_{2d} (4,0); 2/m-C_{2h} (8,0)
I4$_1$/acd-D_{4h}^{20}	142	$\bar{4}$-S_4 (8,0); 222-D_2 (8,0); $\bar{1}$-C_i (16,0)
P$_3$-C_3^1	143	3-C_3 (1,1)
P3$_1$-C_3^2	144	1-C_1 (3,3)
P3$_2$-C_3^3	145	1-C_1 (3,3)
R3-C_3^4	146	3-C_3 (3,1)
P$\bar{3}$-C_{3i}^1	147	$\bar{3}$-C_{3i} (1,0)
R$\bar{3}$-C_{3i}^2	148	$\bar{3}$-C_{3i} (3,0)
P312-D_3^1	149	32-D_3 (1,0)
P321-D_3^2	150	32-D_3 (1,0)
P3$_1$1-D_3^3	151	2-C_2 (3,1)
P3$_1$21-D_3^4	152	2-C_2 (3,1)
P3$_2$12-D_3^5	153	2-C_2 (3,1)
P3$_2$21-D_3^6	154	2-C_2 (3,1)
R32-D_3^7	155	32-D_3 (3,0)
P3m1-C_{3v}^1	156	3m-C_{3v} (1,1)
P31m-C_{3v}^2	157	3m-C_{3v} (1,1)
P3c1-C_{3v}^3	158	3-C_3 (2,1)
P31c-C_{3v}^4	159	3-C_3 (2,1)
R3m-C_{3v}^5	160	3m-C_{3v} (3,1)
R3c-C_{3v}^6	161	3-C_3 (6,1)
P$\bar{3}$1m-D_{3d}^1	162	$\bar{3}$m-D_{3d} (1,0)
P$\bar{3}$1c-D_{3d}^2	163	32-D_3 (2,0); $\bar{3}$-C_{3i} (2,0)
P$\bar{3}$m1-D_{3d}^3	164	$\bar{3}$m-D_{3d} (1,0)
P$\bar{3}$c1-D_{3d}^4	165	32-D_3 (2,0); $\bar{3}$-C_{3i} (2,0)
R$\bar{3}$m-D_{3d}^5	166	$\bar{3}$m-D_{3d} (3,0)
R$\bar{3}$c-D_{3d}^6	167	32-D_3 (6,0); $\bar{3}$-C_{3i} (6,0)
P6-C_6^1	168	6-C_6 (1,1)
P6$_1$-C_6^2	169	1-C_1 (6,3)
P6$_5$-C_6^3	170	1-C_1 (6,3)

Table 3.1.2 (continued)

Space group symbol	No.	Main groups
$P6_2$-C_6^4	171	2-C_2 (3,1)
$P6_4$-C_6^5	172	2-C_2 (3,1)
$P6_3$-C_6^6	173	3-C_3 (2,1)
P6-C_{3h}^1	174	$\bar{6}$-C_{3h} (1,0)
P6/m-C_{6h}^1	175	6/m-C_{6h} (1,0)
$P6_3$/m-C_{6h}^2	176	$\bar{6}$-C_{3h} (2,0); $\bar{3}$-C_{3i} (2,0)
P622-D_6^1	177	622-D_6 (1,0)
$P6_1$22-D_6^2	178	2-C_2 (6,1)
$P6_5$22-D_6^3	179	2-C_2 (6,1)
$P6_2$22-D_6^4	180	222-D_2 (3,0)
$P6_4$22-D_6^5	181	222-D_2 (3,0)
$P6_3$22-D_6^6	182	32-D_3 (2,0)
P6mm-C_{6v}^1	183	6mm-C_{6v} (1,0)
P6cc-C_{6v}^2	184	6-C_6 (2,1)
$P6_3$cm-C_{6v}^3	185	3m-C_{3v} (2,0)
$P6_3$mc-C_{6v}^4	186	3m-C_{3v} (2,0)
P$\bar{6}$m2-D_{3h}^1	187	$\bar{6}$2m-D_{3h} (1,0)
P$3_1$1-D_3^3	188	32-D_3 (2,0); $\bar{6}$-C_{3h} (2,0)
P$\bar{6}$2m-D_{3h}^3	189	$\bar{6}$2m-D_{3h} (1,0)
P$\bar{6}$2c-D_{3h}^4	190	32-D_3 (2,0); $\bar{6}$-C_{3h} (2,0)
P6/mmm-D_{6h}^1	191	6/mmm-D_{6h} (1,0)
P6/mcc-D_{6h}^2	192	622-D_6 (2,0); 6/m-C_{6h} (2,0); 32-D_3 (4,0)
$P6_3$/mcm-D_{6h}^3	193	$\bar{6}$2m-D_{3h} (2,0); $\bar{3}$m-D_{3d} (2,0)
$P6_3$/mmc-D_{6h}^4	194	$\bar{3}$m-D_{3d} (2,0); $\bar{6}$2m-D_{3h} (2,0)
P23-T^1	195	23-T (1,0); 3-C_3 (4,1)
F23-T^2	196	23-T (4,0); 3-C_3 (16,1)
I23-T^3	197	23-T (2,0); 3-C_3 (8,1)
$P2_1$3-T^4	198	3-C_3 (4,1)
$I2_1$3-T^5	199	3-C_3 (8,1); 2-C_2 (12,1)
Pm$\bar{3}$-T_h^1	200	$\bar{3}$m-D_{3d} (1,0); mmm-D_{2h} (3,0)
Pn$\bar{3}$-T_h^2	201	23-T (2,0); $\bar{3}$-C_{3i} (4,0)
Fm$\bar{3}$-T_h^3	202	$\bar{3}$m-D_{3d} (4,0); 23-T (8,0)
Fd$\bar{3}$-T_h^4	203	23-T (8,0); $\bar{3}$-C_{3i} (16,0)
Im$\bar{3}$-T_h^5	204	$\bar{3}$m-D_{3d} (2,0); mmm-D_{2h} (6,0)
Pa$\bar{3}$-T_h^6	205	$\bar{3}$-C_{3i} (4,0)
Ia$\bar{3}$-T_h^7	206	$\bar{3}$-C_{3i} (8,0); 2-C_2 (24,0)
P432-O^1	207	432-O (1,0)
$P4_2$32-O^2	208	23-T (2,0); 32-D_3 (4,0)
F432-O^3	209	432-O (4,0)
$F4_1$32-O^4	210	23-T (8,0); 32-D_3 (16,0)
I432-O^5	211	432-O (2,0)
$P4_3$32-O^6	212	32-D_3 (4,0)
$P4_1$32-O^7	213	32-D_3 (4,0)

Space group symbol	No.	Main groups
$I4_132$-O^8	214	32-D_3 (8,0); 222-D_2 (12,0)
$P\bar{4}3m$-T_d^1	215	$\bar{4}3m$-T_d (1,0)
$F\bar{4}3m$-T_d^2	216	$\bar{4}3m$-T_d (4,0)
$I\bar{4}3m$-T_d^3	217	$\bar{4}3m$-T_d (2,0)
$P\bar{4}3n$-T_d^4	218	23-T (2,0); 222-D_2 (6,0); $\bar{4}$-S_4 (6,0); 3-C_3 (8,1)
$F\bar{4}3c$-T_d^5	219	23-T (8,1); $\bar{4}$-S_4 (24,0); 3-C_3 (32,0)
$I\bar{4}3d$-T_d^6	220	$\bar{4}$-S_4 (12,0); 3-C_3 (16,1)
$Pm\bar{3}m$-O_h^1	221	$m\bar{3}m$-O_h (1,0)
$Pn\bar{3}n$-O_h^2	222	432-O (2,0); $\bar{3}$-C_{3i} (8,0); $\bar{4}$-S_4 (12,0)
$Pm\bar{3}n$-O_h^3	223	$\bar{3}m$-D_{3d} (2,0); mmm-D_{2h} (6,0); $\bar{4}2m$-D_{2d} (6,0)
$Pn\bar{3}m$-O_h^4	224	$\bar{4}3m$-T_d (2,0); $\bar{3}m$-D_{3d} (4,0)
$Fm\bar{3}m$-O_h^5	225	$m\bar{3}m$-O_h (4,0)
$Fm\bar{3}c$-O_h^6	226	432-O (8,0); $\bar{3}m$-D_{3d} (8,0); $\bar{4}2m$-D_{2d} (24,0); 4/m-C_{4h} (24,0)
$Fd\bar{3}m$-O_h^7	227	$\bar{4}3m$-T_d (8,0); $\bar{3}m$-D_{3d} (16,0)
$Fd\bar{3}c$-O_h^8	228	23-T (16,0); 32-D_3 (32,0); $\bar{3}$-C_{3i} (32,0); $\bar{4}$-S_4 (48,0)
$Im\bar{3}m$-O_h^9	229	$m\bar{3}m$-O_h (2,0)
$Ia\bar{3}d$-O_h^{10}	230	$\bar{3}$-C_{3i} (16,0); 32-D_3 (16,0); 222-D_2 (24,0); $\bar{4}$-S_4 (24,0)

For each space group there exists a point group that is not contained in any other point group of the same space group. A group of these properties is called the *main point group* of the space group [5]. The main groups of all 230 space groups are listed in Table 3.1.2.

In connection with irreducible representations of point groups (Chap. 3.4) it is useful to accept the concept of the class of symmetry operations. In the point group S under consideration, operation A is conjugate to operation B under operation X, when the relation holds

$$A = X^{-1}BX$$

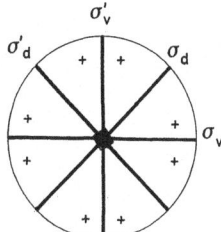

Fig. 3.1.2. Classes of symmetry operations of point group 4 mm-C_{4v}. A stereographic projection.

Table 3.1.3. Classes of symmetry operations of the point groups

Crystal system	Point group	Classes of operations[1]	Order of group
Triclinic	$1\text{-}C_1$	E	1
	$\bar{1}\text{-}C_i$	$E\ i$	2
Monoclinic	$2\text{-}C_2$	$E\ C_2$	2
	$m\text{-}C_s$	$E\ \sigma_h$	2
	$2/m\text{-}C_{2h}$	$E\ C_2\ i\ \sigma_h$	4
Orthorhombic	$222\text{-}D_2$	$E\ C_2\ C_2'\ C_2''$	4
	$mm2\text{-}C_{2v}$	$E\ C_2\ \sigma_v\ \sigma_v'$	4
	$mmm\text{-}D_{2h}$	$E\ C_2\ C_2'\ C_2''\ i\ \sigma'\ \sigma''$	8
Tetragonal	$4\text{-}C_4$	$E\ C_4\ C_2\ C_4^3$	4
	$\bar{4}\text{-}S_4$	$E\ S_4\ C_2\ S_4^3$	4
	$4/m\text{-}C_{4h}$	$E\ C_4\ C_2\ C_4^3\ i\ S_4^3\ \sigma_h\ S_4$	8
	$422\text{-}D_4$	$E\ 2C_4\ C_2\ 2C_2'\ 2C_2''$	8
	$4mm\text{-}C_{4v}$	$E\ 2C_4\ C_2\ 2\sigma_v\ 2\sigma_d$	8
	$\bar{4}2m\text{-}D_{2d}$	$E\ 2S_4\ C_2\ 2C_2'\ 2\sigma_d$	8
	$4/mmm\text{-}D_{4h}$	$E\ 2C_4\ C_2\ 2C_2'\ 2C_2''\ i\ 2S_4\ \sigma_h\ 2\sigma_v\ 2\sigma_d$	16
Trigonal	$3\text{-}C_3$	$E\ C_3\ C_3^2$	3
	$\bar{3}\text{-}C_{3i}$	$E\ C_3\ C_3^2\ i\ S_6^5\ S_6$	6
	$32\text{-}D_3$	$E\ 2C_3\ 3C_2'$	6
	$3m\text{-}C_{3v}$	$E\ 2C_3\ 3\sigma_v$	6
	$\bar{3}m\text{-}D_{3d}$	$E\ 2C_3\ 3C_2'\ i\ 2S_6\ 3\sigma_d$	12
Hexagonal	$6\text{-}C_6$	$E\ C_6\ C_3\ C_2\ C_3^2\ C_6^5$	6
	$\bar{6}\text{-}C_{3h}$	$E\ C_3\ C_3^2\ \sigma_h\ S_3\ S_3^5$	6
	$6/m\text{-}C_{6h}$	$E\ C_6\ C_3\ C_2\ C_3^2\ C_6^5\ i\ S_3^5\ S_6^5\ \sigma_h\ S_6\ S_3$	12
	$622\text{-}D_6$	$E\ 2C_6\ 2C_3\ C_2\ 2C_2'\ 3C_2''$	12
	$6mm\text{-}C_{6v}$	$E\ 2C_6\ 2C_3\ C_2\ 3\sigma_v\ 3\sigma_d$	12
	$\bar{6}m2\text{-}D_{3h}$	$E\ 2C_3\ 3C_2'\ \sigma_h\ 2S_3\ 3\sigma_v$	12
	$6/mmm\text{-}D_{6h}$	$E\ 2C_6\ 2C_3\ C_2\ 3C_2'\ 3C_2''\ i\ 2S_3\ 2S_6$ $\sigma_h\ 3\sigma_d\ 3\sigma_v$	24
Cubic	$23\text{-}T$	$E\ 4C_3\ 4C_3^2\ 3C_2$	12
	$m\bar{3}\text{-}T_h$	$E\ 4C_3\ 4C_3^2\ 3C_2\ i\ 4S_6^5\ 4S_6\ 3\sigma_h$	24
	$432\text{-}O$	$E\ 8C_3\ 6C_2'\ 6C_4\ 3C_2$	24
	$\bar{4}3m\text{-}T_d$	$E\ 8C_3\ 3C_2\ 6S_4\ 6\sigma_d$	24
	$m\bar{3}m\text{-}O_h$	$E\ 8C_3\ 6C_2\ 6C_4\ 3C_2\ i\ 6S_4\ 8S_6\ 3\sigma_h$ $6\sigma_d$	48

[a] E is identity operation.
σ_n, σ_v, σ_d and σ are mirror planes.
Number of rotation operations by axes C_n or S_n ($S_1 \equiv \sigma$) is expressed by integer for appropriate symbol.

where X^{-1} is the inverse operation to that of X, i.e. $XX^{-1} = E$ (E being the identity operation). Because of the properties of conjugate elements of the group (see Chap. 3.2), the point group S decomposes to subsets of the conjugate operations called *classes of symmetry operations*. Operations from different symmetry elements, being symmetrically equivalent, belong to the same operation class. For example, two mutually normal symmetry planes of the point group 4 mm-C_{4v} are symmetrically equivalent. Thus their symmetry operations σ_v, σ_v' and σ_d, σ_d (Fig. 3.1.2) form two classes, $2\sigma_v$ and $2\sigma_d$. Table 3.1.3 brings a list of classes contained in point groups.

3.2 Lattice Complexes

From a geometrical view crystal structures may also be described on the basis of crystallographically equivalent monatomic entities, of which the structure consists. Thus, for example, the structure of CsCl may be described as the mutual penetration of a set of Cs^+ and a set of Cl^- entities. The first set consists of regularly arranged entities in a space with cubic unit cell. The sites of Cs^+ and Cl^- entities in this unit cell may be expressed by fraction coordinates 0, 0, 0 and 1/2, 1/2, 1/2, or in eights of the parameter of the unit cell 0, 0, 0 and 4, 4, 4. Niggli [6] called these sets lattice complexes. This concept was modified by Hermann [7] and extended by Donnay, Hallner and Niggli [8]. More precise definitions of lattice complex concept were formulated in [9—11]. Lattice complexes are also described in *International Tables for Crystallography* [12] and in [3]. Some crystallochemical applications of lattice complexes concepts are described in [13–17].

The lattice points, called *nodes* may be expressed as terminal points of radius vectors $r_{u, v, w}$, given by Eq. (1.2.1). In first approximation the set of crystallographically equivalent points of space E^3 may be considered as lattice complex. The array of Cs^+ and Cl^- entities in the structure of CsCl may then be described by a cubic lattice complex marked with the symbol P. Similarly the array of Na^+ and Cl^- entities in the structure of NaCl may be described by lattice complex F.

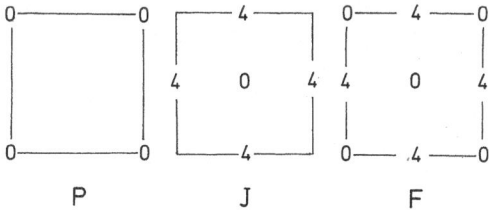

Fig. 3.2.1. Some cubic lattice complexes [3]. Hights in eights of the unit cell parameter.

Cubic lattice complexes P, I and F are graphically shown in Fig. 3.2.1. If the different metric of structures is not taken into consideration, then some other structures can be described by means of the below cubic lattice complexes:

CaF_2	$F; P$
ZnS (Zinkblende)	$F; F$
NaCl	$F; F$

BiF_3 (BiFF) $F; F; P$
$AuCu_3$ (ordered) $P; J$
ReO_3 $P; J$
$CaTiO_3$ $P; P; J$
NbO $J; J$
U_4S_3 ($U_1U_3S_3$) $P; J; J$

Such a description of structures is independent of the choice of crystallographic origin and space group. The above mentioned cubic lattice complexes may have their symmetry described by some cubic space group. For a certain lattice complex there is only one of them showing the greatest point symmetry. Thus the lattice complex P can have its symmetry expressed by one of the space groups P$\bar{4}$3m, Ia3, I432, Im3m, P$\bar{4}$3m, Pm3, Pm3m, Fm3c, P23, P$\bar{4}$3m, P432, F$\bar{4}$3c however, only in the space group Pm3m is îts maximum point symmetry m$\bar{3}$m-O_h. Similarly the lattice complex F can have a symmetry with the greatest point symmetry of (m$\bar{3}$m-O_h) described by group Fm3m.

The conception of cubic lattice complexes was used as a classification basis for cubic structure types by several authors [13–17]. However, the crystal structure of coordination compounds for the major part show lower symmetries. In order to classify the relative positions of their central monatomic entities a more exact definition of lattice complexes and a more detailed analysis of their properties is necessary.

The symmetry of crystal structure can be described by some of 230 space groups G. Its subgroups G_1, G_2, G_3, ... G_n are conjugate, when for any arbitrary couple G_i, G_j there exists at least one symmetry operation of group G transforming G_i into G_j. If subgroup G_i lets the point set of space E^3 become invariant, then this subgroup becomes a *site symmetry group*. It is one of the 32 symmetry point groups. A space point is a point of *general position*, when its symmetry point group is 1-C_1. In the other cases it has a *special position* with respect to group G. The set of points, for which G_i is a conjugate subgroup with respect to G, is called the *Wickoff position*. In *International Tables for Crystallography* are the *Wickoff positions* of space groups marked with letters called the *Wickoff letters* [18] or *Wickoff notation* [19]. The rank of group G_i is a *multiplicity of the Wickoff position*. Thus space group P2/c (No. 13) shows the Wickoff positions:

Multiplicity	Wickoff letter	Site symmetry	Coordinates of equivalent positions
4	g	1-C_1	$x, y, z; \bar{x}, y, \bar{z} + \frac{1}{2}; \bar{x}, \bar{y}, \bar{z}; x, \bar{y}, z + \frac{1}{2}$
2	f	2-C_2	$\frac{1}{2}, y, \frac{1}{4}; \frac{1}{2}, \bar{y}, \frac{3}{4}$
2	e	2-C_2	$0, y, \frac{1}{4}; 0, \bar{y}, \frac{3}{4}$

Multiplicity	Wickoff letter	Site symmetry	Coordinates of equivalent positions
2	d	$\bar{1}\text{-}C_i$	$\frac{1}{2}, 0, 0; \frac{1}{2}, 0, \frac{1}{2}$
2	c	$\bar{1}\text{-}C_i$	$0, \frac{1}{2}, 0; 0, \frac{1}{2}, \frac{1}{2}$
2	b	$\bar{1}\text{-}C_i$	$\frac{1}{2}, \frac{1}{2}, 0; \frac{1}{2}, \frac{1}{2}, \frac{1}{2}$
2	a	$\bar{1}\text{-}C_i$	$0, 0, 0; 0, 0, \frac{1}{2}$

Positions a, b, c and d are symmetrically equivalent. They are the centres of symmetry which are crystallographically indistinguishable. Their notations may be cyclically permuted a–d. Thus they may be subject to permutation under isomorphic mapping of group P2/c into themselves. Otherwise stated, Wickoff positions a, b, c, d can be permuted under *automorphism* of group P2/c. Similarly Wickoff positions e and f may be permuted under automorphism of group P2/c. The set of Wickoff positions of group G, which can be permuted under its automorphism, is the *Wickoff set*. Apparently in our example, Wickoff position g is by itself a Wickoff set. The set of symmetrically equivalent points of Wickoff position is a *point configuration*. In the case of our group P2/c a point set in the given coordinate system is expressed by coordinates $x, y, z; \bar{x}, y, z + \frac{1}{2}; \bar{x}, \bar{y}, \bar{z}; x, \bar{y}, z + \frac{1}{2}$ and the point configuration of Wickoff position is g. It is apparent that each space group has indefinitely many point configurations. The Wickoff set a, b, c, d, of group P2/c is a system of points in space E^3 being topologically congruent with Wickoff sets a, b, c, ... , h in group P2/m with the fractional coordinates $(0, 0, 0)$; $\left(0, \frac{1}{2}, 0\right)$; $\left(0, 0, \frac{1}{2}\right)$ ... ; $\left(\frac{1}{2}, \frac{1}{2}, \frac{1}{2}\right)$. Congruent or enantiomorph Wickoff sets are sets of the same type. The set of all point configurations, which can be generated by one type of Wickoff sets is a *lattice complex*.

Crystal lattices can have seven symmetries, which can be expressed by point groups $\bar{1}\text{-}C_i$, $2/\text{m}\text{-}C_{2h}$, $\text{mmm}\text{-}D_{2h}$, $4/\text{mmm}\text{-}D_{4h}$, $\bar{3}\text{m}\text{-}D_{3d}$, $6/\text{mmm}\text{-}D_{6h}$ and $\text{m3m}\text{-}O_h$. The seven crystal systems belonging to these groups are: triclinic, monoclinic, orthorhombic, tetragonal, trigonal, hexagonal and cubic systems. From the above given definition of a lattice complex the following consequences may be deduced:

(a) a lattice complex may be realized in several space groups;
(b) all positions in which a lattice complex is realized belong to the same crystal system

In the presented example of space group P2/m a monoclinic lattice complex P is realized (Fig. 3.2.2) in the system of sites a, b, c, ... , h of the site symmetry $2/m$-C_{2h}. In Table 3.2.1 all space groups are listed in which this lattice complex can be realized. The sites of the lattice complex, being realized in a certain space group or in different space groups, can exhibit different site symmetries and they can also be differently oriented. We say that a lattice complex can be realized in different *representations*; however, only one of them shows the highest site symmetry. This one is called the *standard representation*. The other representations alternate. The space group in which the lattice complex is realized in standard representation is called the *characteristic space group*. Thus the monoclinic complex P can occur in representations (Table 3.2.1), from which only one shows its highest

Table 3.2.1. Realisation of monoclinic lattice complex P in the different space groups

Space group (Wickoff letter)	Site symmetry	Multiplicity	Designation of lattice complex representation
P2/m (a, b, c, d, e, f, g, h)	$2/m$-C_{2h}	1	*(i)*
Pm (a, b)	m-C_s	1	*(i)*
P2 (a, b, c, d)	2-C_2	1	*(i)*
P2/m (a, b, c, d)	$\bar{1}$-C_i	2	*(ii)*
P2/c (a, b, c, d)	$\bar{1}$-C_i	2	*(iii)*
C2/m (e, f)	$\bar{1}$-C_i	4	*(iv)*

site symmetry $(2/m$-$C_{2h})$. Thus the characteristic space group of the monoclinic lattice complex P is P2/m. Representations of this lattice complex are shown in Fig. 3.2.2. The number of equivalent points in the unit cell, i.e. the multiplicity of the lattice complex realization, apparently depends on the space group and the choice of the unit cell. The number of equivalent points contained in a conventionally elected unit cell of the standard representation of a lattice complex is its *multiplicity*.

The topological properties of all point configurations may be studied within 402 lattice complexes tabulated in *International Tables for Crystallography* [12]. The point configurations of the lattice complex, in a certain representation consisting of points of one Wickoff position, occupy the same point position. Thus the point configurations of a monoclinic lattice complex P in representation number *(ii)* (Table 3.2.1) described by fraction coordinates x, 0, z and x, $\frac{1}{2}$, z occupy the point positions a and b. Their site symmetry is m-C_s, so that they occupy positions on the mirror planes of the space group Pm. The independent variation of parameters x, z leads to point configurations being, however, equivalent on shifting. The number of independent coordinate parameters is equal for all positions of the same lattice complex. The number of these parameters is equal to the *degrees of freedom of the lattice complex*. Degrees of freedom of the point position means the dimension of the point set of one site symmetry. In many cases the degrees of freedom of all point positions is the same as the degrees of freedom of their lattice complex. However, our example shows that this does not hold generally. The point positions of (*ii*) and (*iii*) (Table 3.2.1) exhibit two degrees of freedom. The monoclinic lattice complex *P*, however, does not exhibit any degree of freedom. Such a lattice complex is called *invariant*. Lattice complexes can have one, two or three degrees of freedom and according to this they are called mono-, bi- and trivariant. The latest lattice complexes are *general*, the other are *special*.

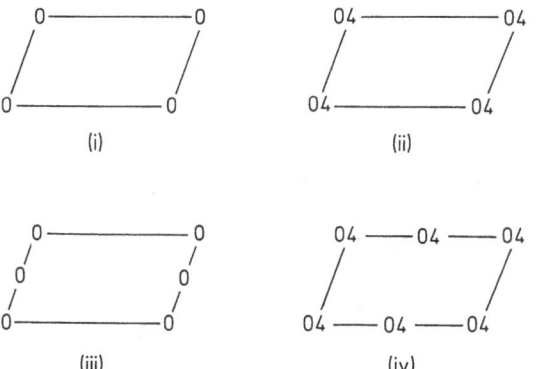

Fig. 3.2.2. Representations of monoclinic lattice complex *P* [3]. Hights in eights of the unit cell parameter *b*.

Two point positions in one space group, or in different space groups, belong to the same lattice complex, if to each point configuration of the first position there exists a congruent or enantiomorph configuration of the second position. An algebraic definition of the point position and the algorithm, by means of which one can assign this position to its lattice complex, is presented in [10]. In [11] a lattice complex is defined using concepts of the set theory and the group theory. From this point of view a lattice complex is defined as a set of point configurations. When lattice complex L_1 contains all point configurations of another complex L_2, i.e. L_2 is a subset of the lattice complex L_1 ($L_2 \subset L_1$), then L_2 is a *limiting lattice complex* of the lattice complex L_1, and conversely, L_2 is a *comprehensive* lattice complex to L_1. Point configurations of the limiting lattice complex L_2 can be generated within complex L_1 by introducing restrictions on positional parameters. Thus, for example, a lattice complex being realized in position *j* of the space group P23 is expressed by positional parameters:

x, y, z; z, x, y; y, z, x; $x, \bar{y}, \bar{z}$; $z, \bar{x}, \bar{y}$; $y, z, \bar{x}$; $\bar{x}, y, \bar{z}$; $\bar{z}, x, \bar{y}$; $\bar{y}, z, \bar{x}$; $\bar{x}, \bar{y}, z$; $\bar{z}, \bar{x}, y$; $\bar{y}, \bar{z}, x$. By introducing restriction $y = x$, $z = 0$, we obtain position *i* of space groups Pm3m and P432:

$x, x, 0$; $0, x, x$; $x, 0, x$; $x, \bar{x}, 0$; $0, \bar{x}, \bar{x}$; $x, 0, \bar{x}$; $\bar{x}, x, 0$; $0, x, \bar{x}$; $\bar{x}, 0, \bar{x}$; $\bar{x}, \bar{x}, 0$; $0, \bar{x}, x$; $\bar{x}, 0, x$. Thus a lattice complex realized in position *i* of space groups Pm3m and P432 is then a limiting lattice complex of the first complex realized in position *j* of space group P23. Both positions *i* and *j* have the same multiplicity. We say that position *j* in space group P23 for special coordinates $y = x$, $z = 0$ *simulates* position *i* of space groups Pm3m and P432 while keeping the multiplicity of the

Table 3.2.2. Invariant lattice complexes [3]

Symbol	Multiplicity	Coordinates of equivalent points[a]
A	2	$0\,0\,0, 0\,\frac{1}{2}\,\frac{1}{2}$
B	2	$0\,0\,0, \frac{1}{2}\,0\,\frac{1}{2}$
C	2	$0\,0\,0, \frac{1}{2}\,\frac{1}{2}\,0$
D	8	$0\,0\,0, \frac{1}{4}\,\frac{1}{4}\,\frac{1}{4}, \left(\frac{1}{2}\,\frac{1}{2}\,0, \frac{3}{4}\,\frac{3}{4}\,\frac{1}{4}\right)\circlearrowright$
E	2	$\frac{1}{3}\,\frac{2}{3}\,\frac{1}{4}, \frac{2}{3}\,\frac{1}{3}\,\frac{3}{4}$
F	4	$0\,0\,0, \frac{1}{2}\,\frac{1}{2}\,0\circlearrowright$
G	2	$\frac{1}{3}\,\frac{2}{3}\,0, \frac{2}{3}\,\frac{1}{3}\,0$ (hexagonal cell)
I	2	$0\,0\,0, \frac{1}{2}\,\frac{1}{2}\,\frac{1}{2}$
J	3	$\frac{1}{2}\,0\,\frac{1}{2}\circlearrowright$
J^*	6	$\left(\frac{1}{2}\,\frac{1}{2}\,0, \frac{1}{2}\,0\,0\right)\circlearrowright$
M	9	$\frac{1}{2}\,0\,0, 0\,\frac{1}{2}\,0, \frac{1}{2}\,\frac{1}{2}\,0, \frac{5}{6}\,\frac{2}{3}\,\frac{2}{3}, \frac{1}{3}\,\frac{1}{6}\,\frac{2}{3}$
		$\frac{5}{6}\,\frac{1}{6}\,\frac{2}{3}, \frac{1}{6}\,\frac{1}{3}\,\frac{1}{3}, \frac{2}{3}\,\frac{5}{6}\,\frac{1}{3}, \frac{1}{6}\,\frac{5}{6}\,\frac{1}{3}$
N	3	$\frac{1}{2}\,0\,0, 0\,\frac{1}{2}\,0, \frac{1}{2}\,\frac{1}{2}\,0$
P	1	$0\,0\,0$
^+Q	3	$\frac{1}{2}\,0\,0, 0\,\frac{1}{2}\,\frac{2}{3}, \frac{1}{2}\,\frac{1}{2}\,\frac{1}{3}$ (hexagonal cell)
R	3	$0\,0\,0, \frac{1}{3}\,\frac{2}{3}\,\frac{2}{3}, \frac{2}{3}\,\frac{1}{3}\,\frac{1}{3}$ (hexagonal cell)
S	12	$\left(0\,\frac{1}{4}\,\frac{3}{8}, \frac{5}{8}\,\frac{1}{2}\,\frac{1}{4}, \frac{3}{4}\,\frac{1}{8}\,0, \frac{1}{2}\,\frac{3}{4}\,\frac{7}{8}\right)\circlearrowright$
S^*	24	$\left(0\,\frac{1}{4}\,\frac{3}{8}, 0\,\frac{3}{4}\,\frac{1}{8}, \frac{1}{2}\,\frac{3}{4}\,\frac{7}{8}, \frac{1}{2}\,\frac{1}{4}\,\frac{5}{8}\right)\circlearrowright$
		$\left(0\,\frac{3}{4}\,\frac{5}{8}, 0\,\frac{1}{4}\,\frac{7}{8}, \frac{1}{2}\,\frac{1}{4}\,\frac{1}{8}, \frac{1}{2}\,\frac{3}{4}\,\frac{3}{8}\right)\circlearrowright$
T	16	$\frac{1}{8}\,\frac{1}{8}\,\frac{1}{8}, \frac{3}{8}\,\frac{1}{8}\,\frac{3}{8}, \frac{5}{8}\,\frac{5}{8}\,\frac{1}{8}\circlearrowright$
		$\left(\frac{7}{8}\,\frac{5}{8}\,\frac{3}{8}, \frac{7}{8}\,\frac{1}{8}\,\frac{7}{8}, \frac{3}{8}\,\frac{5}{8}\,\frac{7}{8}\right)\circlearrowright$
^+V	12	$\left(\frac{1}{4}\,\frac{1}{8}\,0, \frac{3}{4}\,\frac{3}{8}\,0, \frac{3}{4}\,\frac{5}{8}\,\frac{1}{2}, \frac{7}{8}\,\frac{1}{2}\,\frac{1}{4}\right)\circlearrowright$

Table 3.2.2 (continued)

Symbol	Multiplicity	Coordinates of equivalent points[a]
W	6	$\left(\dfrac{1}{2}\dfrac{1}{4}\,0,\,0\,\dfrac{1}{2}\dfrac{3}{4}\right)\circlearrowright$
$W*$	12	$\left(0\,\dfrac{1}{2}\dfrac{1}{4},\,0\,\dfrac{1}{2}\dfrac{3}{4},\,\dfrac{1}{2}\,0\,\dfrac{3}{4},\,\dfrac{1}{2}\,0\,\dfrac{1}{4}\right)\circlearrowright$
^{+}Y	4	$\dfrac{1}{8}\dfrac{1}{8}\dfrac{1}{8},\,\dfrac{7}{8}\dfrac{5}{8}\dfrac{3}{8}\circlearrowright$
$^{+}Y*$	8	$\dfrac{1}{8}\dfrac{1}{8}\dfrac{1}{8},\,\dfrac{7}{8}\dfrac{5}{8}\dfrac{3}{8}\circlearrowright,\,\dfrac{5}{8}\dfrac{5}{8}\dfrac{5}{8},\,\dfrac{3}{8}\dfrac{1}{8}\dfrac{7}{8}\circlearrowright$
$Y**$	16	$\dfrac{1}{8}\dfrac{1}{8}\dfrac{1}{8},\,\dfrac{7}{8}\dfrac{5}{8}\dfrac{3}{8}\circlearrowright,\,\dfrac{5}{8}\dfrac{5}{8}\dfrac{5}{8},\,\dfrac{3}{8}\dfrac{1}{8}\dfrac{7}{8}\circlearrowright,$ $\dfrac{7}{8}\dfrac{7}{8}\dfrac{7}{8},\,\dfrac{1}{8}\dfrac{3}{8}\dfrac{5}{8}\circlearrowright,\,\dfrac{3}{8}\dfrac{3}{8}\dfrac{3}{8},\,\dfrac{5}{8}\dfrac{7}{8}\dfrac{1}{8}\circlearrowright$

[a] Symbol $\circlearrowright$ indicates the three cyclic parmutations of the proceeding triplet or of more triplets enclosed between parantheses.

lattice complex. A limiting lattice complex generated by this way shows, however, less degrees of freedom against its comprehensive lattice complex. The relationship between a limiting lattice complex and comprehensive lattice complexes can be the basis for a classification of crystal structures.

The symbolics of lattice complexes [3] are based on their classification using the described splitting of their points, as also on their topological properties. In Table 3.2.2 symbols of invariant lattice complexes are presented. For a concrete point configuration being contained in several lattice complexes it appears reasonable to find the least comprehensive lattice complex. One of the possibilities, how to deduce it, is the above discussed restriction on the coordinate parameters. Another way would be a restriction on metric, or a combination of both methods.

For our purposes, however, it is more suitable to use such a classification of lattice complexes, that starts from certain lattice complexes. Their points produce by splitting the other complexes. The starting lattice complexes are called generating lattice complexes. They can be invariant or limiting lattice complexes. By the splitting of their points variant complexes are produced.

The choice of the generating lattice complex, using splitting by which we can obtain the given variant complex, is not always an unambiguous problem. To the given variant complex it can be a limiting lattice complex, or an invariant lattice complex. As Table 3.1.2 shows, each space group covers at least one main symmetry group, to which a point position corresponds. The lattice complex, generated from it by means of operations of the symmetry group, is called "Hauptgitter" according to Weissenberg [5] and according to proposal in [3] it is the *Weissenberg lattice complex*. Altogether there are 67 Weissenberg lattice complexes. By splitting we can deduce from them 402 lattice complexes. Weissenberg lattice complexes are tabulated in [3]. They consist of variant lattice complexes and of some univariant, bivariant and trivariant lattice complexes. Except for the trivariant ones, all of them can be used as generating lattice complexes. As it follows from the second relation (3.1.1), their multiplicity for the given space group is minimum, *i.e.* their multiplicity never falls for any special values of coordinates.

3.3 Domains of Influence and Lattice Complexes

In Chapter 1.2 we introduced the concept of domain of influence, while we did not consider the symmetry of discontinuum, except the translation symmetry of its lattice. We supposed from this the symmetry of each site $1\text{-}C_1$, i.e. generally the *heterogeneous set of points* in the respective unit cell. In this case the multiplicity m of each site is unity, if the conventionally chosen unit cell is P or R, it is two, if the unit cell is I, A or C and it is four, if the cell is F. In concrete cases the sites in the unit cell can be symmetrically equivalent. For such a homogeneous set of points the influence domain of a certain site A covers all the points, being nearer to site A, than any other of its equivalent site. The concept of the influence domain introduced in this way applies to sites of a certain lattice complex [20, 21]. For all symmetrically equivalent sites A_i (centres of influence domains) these domains are congruent or mirror congruent. Equally as centres A_i also these domains are symmetrically equivalent.

In general: For different crystallographic sites these domains are different and they are also limited by convex polyhedra. One can, however, delimit such sets of centres, within which these polyhedra are of the same type, *i.e.* they have the same number of faces and edges. They may only differ by their sizes. Figure 3.3.1 shows

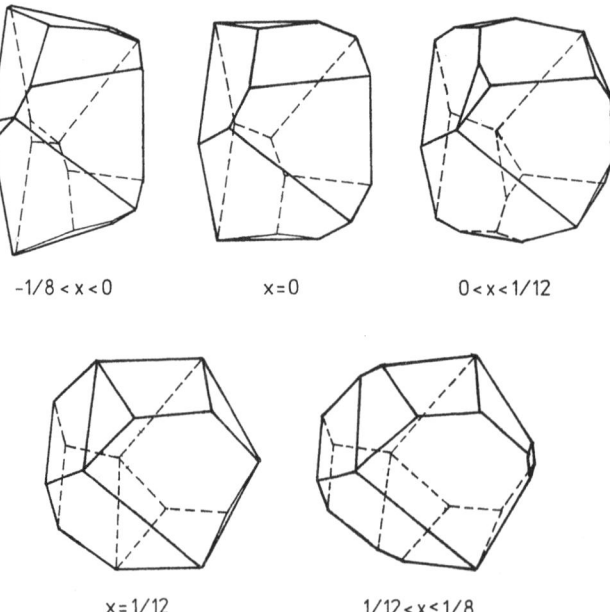

-1/8 < x < 0   x = 0   0 < x < 1/12

x = 1/12   1/12 < x ≤ 1/8

Fig. 3.3.1. Influence domains of univariant lattice complex $I\bar{4}3d(d)$ $x, 0, \dfrac{1}{4}$ [21]. A homogeneous set of points.

types of polyhedra of the influence domains of univariant cubic lattice complex .3.S2z, occurring in position d of the space group I$\bar{4}$3d. In [21] this kind of polyhedra are discussed for all cubic invariant, univariant and bivariant lattice complexes. The volume of polyhedron containing an influence domain with multiplicity of its centre m_A is $V_A = V/m_A$, where V is the volume of the unit cell. The multiplicity of Weissenberg lattice complex realized in a certain space group is minimum, compared with other lattice complexes of the same space group; in this way the influence domain of Weissenberg complex will have the maximum volume.

3.4 Matrix Representations of Symmetry Point Groups of Crystallographic Sites

Site symmetry fulfils a significant role in solving quantum mechanical problems of particles occupying a crystallographic site set. The Schrödinger equation of such a system may be written as an operator equation:

$$\mathbb{H}\Psi_i = E_i\Psi_i \tag{3.4.1}$$

$\mathbb{H}$ is Hamiltonian, which for a system of p particles can be written in the classic form

$$\mathbb{H} = \frac{1}{2}\sum_{i=1}^{p} m_i v_i^2 + U(r_1, r_2, r_3, \ldots, r_p)$$

where v_i is the velocity of ith particle of mass m_i and $U(r_1, r_2, r_3 \ldots r_p)$ means the potential energy. For the case of *non-degenerate* state to each eigenvalue (energy) E_i in Eq. (3.4.1) corresponds one, and only one eigenfunction (wave function). Let the potential energy of such a system be invariant with respect to symmetry operation expressed by operator $\mathbb{R}$, which can be expressed by equation:

$$\mathbb{R}\,U(r_1, r_2, r_3, \ldots, r_p) = U(r_1, r_2, r_3, \ldots, r_p)$$

Similarly for the kinetic energy as a component of Hamiltonian the relation holds[1]

$$\mathbb{R}\cdot\frac{1}{2}\left(\sum_{i=1}^{p} m_i v_i^2\right) = \frac{1}{2}\sum_{i=1}^{p} m_i \mathbb{R}\dot{r}_i^2 = \frac{1}{2}\sum_{i=1}^{p} m_i \dot{r}_i^2$$

Thus Hamiltonian of the Schrödinger Eq. (3.4.1) is also *invariant* with respect to site symmetry operations. Since E_i are numerical constants and Hamiltonian shows such a property, one can write for a non-degenerate case:

$$\mathbb{R}\,\mathbb{H}\Psi_i = \mathbb{R}E_i\Psi_i$$

$$\mathbb{H}(\mathbb{R}\Psi_i) = E_i(\mathbb{R}\Psi_i)$$

---

[1] $\dot{r}_i$ is time derivative r_i

Thus also $\mathbb{R}\,\Psi_i$ are eigenfunctions of operator $\mathbb{H}$. Let us now consider another property of operator $\mathbb{R}$. In a vector space it transforms an arbitrary vector r in another vector r' of the same vector space:

$$\mathbb{R}r = r' \tag{3.4.2}$$

One can easily make sure that for two arbitrary vectors r_1 and r_2 the relation holds

$$\mathbb{R}(c_1 r_1 + c_2 r_2) = c_1 \mathbb{R}r_1 + c_2 \mathbb{R}r_2$$

where c_1 and c_2 are constants. Thus it is a *linear operator*. Let $e_1, e_2, e_3, \dots, e_n$ be unit linear independent vectors of this vector space, called basis of n-dimensional space M. Then, according to inferences of vector algebra, each vector may be expressed as their linear combination:

$$r = \sum_{i=1}^{n} \alpha_i e_i$$

where α_i are constants. Inserting this development into Eq. (3.4.2), we obtain

$$r' = \mathbb{R}r = \mathbb{R}\sum_{i=1}^{n} \alpha_i e_i = \sum_{i=1}^{n} \alpha_i \mathbb{R}e_i = \sum_{i=1}^{n} \alpha_i \mathbb{R}e_i = \sum_{i=1}^{n} \alpha_i e_i'$$

Thus operator $\mathbb{R}$ transforms not only vector r to r', but also *basis* $[e]$ to $[e']$. This transformation can also be expressed by formula

$$e_i' = \sum_{j=1}^{n} \Gamma_{ij}' e_j$$

where Γ_{ij}' are constants. This equation can be written using matrix notation:

$$
\begin{bmatrix} e_1' \\ e_2' \\ e_3' \\ \vdots \\ e_n' \end{bmatrix}
=
\begin{bmatrix}
\Gamma_{11}' & \Gamma_{12}' & \Gamma_{13}' & \cdots & \Gamma_{1n}' \\
\Gamma_{21}' & \Gamma_{22}' & \Gamma_{23}' & \cdots & \Gamma_{2n}' \\
\Gamma_{31}' & \Gamma_{32}' & \Gamma_{33}' & \cdots & \Gamma_{3n}' \\
\vdots & & & & \vdots \\
\Gamma_{n1}' & \Gamma_{n2}' & \Gamma_{n3}' & \cdots & \Gamma_{nn}'
\end{bmatrix}
\begin{bmatrix} e_1 \\ e_2 \\ e_3 \\ \vdots \\ e_n \end{bmatrix}
$$

or

$$[e'] = \Gamma'[e] \tag{3.4.3}$$

Thus to each symmetry operator $\mathbb{R}$, a square matrix corresponds, performing one basis into another of the same vector space. For operators and matrices corresponding to them the rules of operator and matrix algebra apply [24].

The eigenvalue form of the Schrödinger Eq. (3.4.1) can be performed for further purposes to a more suitable form, covering also the *degenerate state*:

$$\mathbb{H}\varphi = v \tag{3.4.4}$$

Wave functions φ and v can be expanded into rows by means of a set of n orthogonal and normalized[1] functions ψ_i of operator $\mathbb{R}$:

$$\varphi = \sum_{j=1}^{n} a_j \psi_j \qquad v = \sum_{j=1}^{n} b_j \psi_j \tag{3.4.5}$$

Inserting developments (3.4.5) into Eq. (3.4.4) gives

$$\mathbb{H}\left(\sum_{j=1}^{n} a_j \psi_j\right) = \sum_{j=1}^{n} b_j \psi_j$$

Using the properties of linear operator $\mathbb{H}$ allows to modify this equation as

$$\sum_{j=1}^{n} a_j \mathbb{H}\psi_j = \sum_{j=1}^{n} b_j \psi_j$$

By multiplying this equation by function ψ_i^* and by its integration through the whole space we obtain equation

$$\sum_{j=1}^{n} a_j \mathrm{H}_{ij} = \sum_{j=1}^{n} b_j \tag{3.4.6}$$

where $\mathrm{H}_{ij} = \int \psi_i^* \mathbb{H}\psi_{ij}\, d\tau$. Equation (3.4.6) again may be expressed using matrix notation

$$HA = B \tag{3.4.7}$$

where

$$H = \begin{bmatrix} \mathbb{H}_{11} & \mathbb{H}_{12} & \mathbb{H}_{13} & \cdots & \mathbb{H}_{1n} \\ \mathbb{H}_{21} & \mathbb{H}_{22} & \mathbb{H}_{23} & \cdots & \mathbb{H}_{2n} \\ \mathbb{H}_{31} & \mathbb{H}_{32} & \mathbb{H}_{33} & \cdots & \mathbb{H}_{3n} \\ \vdots & & & & \vdots \\ \mathbb{H}_{n1} & \mathbb{H}_{n2} & \mathbb{H}_{n3} & \cdots & \mathbb{H}_{nn} \end{bmatrix} \quad A = \begin{bmatrix} a_1 \\ a_2 \\ a_3 \\ \vdots \\ a_n \end{bmatrix} \quad B = \begin{bmatrix} b_1 \\ b_2 \\ b_3 \\ \vdots \\ b_n \end{bmatrix}$$

Now we have obtained the *matrix form* of Schrödinger equation. Each element **g** of the point group **S** corresponds according to Eq. (3.4.3) matrix $\Gamma'(\mathbf{g})$. For the

---

[1] Functions ψ_i and ψ_j are orthogonal and normalized, when it holds $\int \psi_i^* \psi_j\, d\tau = \delta_{ij}$; ψ_i^* is a complex conjugate function to ψ_i and δ_{ij} is Kronecker symbol *i.e.* $\delta_{ij} = 1$ for $i = j$ and $\delta_{ij} = 0$ for $i \neq j$.

product of elements g_1 and g_2 of group S it further holds: $\Gamma'(g_1 g_2) = \Gamma'(g_1)\, \Gamma'(g_2)$
Using this property of matrices Γ' one can prove that they form a group (Appendix D).
Such a *homomorphic* image of point group onto a group of matrices is called *matrix
representation* of point group S. The dimension of space in which matrices
are defined is the dimension of matrix representation. The column matrices A and B of
the Schrödinger equation form (3.4.7) show the same dimension. Table 3.4.1 brings

Table 3.4.1. Multiplication table of point group
$2/m\text{-}C_{2h}$

C_{2h}	E	C_2^1	i	σ_h
E	E	C_2^1	i	σ_h
C_2^1	C_2^1	E	σ_h	i
i	i	σ_h	E	C_2^1
σ_h	σ_h	i	C_2^1	E

Matrices of symmetry operation

$$E \equiv \begin{bmatrix} 1 & 0 & 0 \\ 0 & 1 & 0 \\ 0 & 0 & 1 \end{bmatrix} \quad C_2^1 \equiv \begin{bmatrix} -1 & 0 & 0 \\ 0 & -1 & 0 \\ 0 & 0 & 1 \end{bmatrix} \quad i \equiv \begin{bmatrix} -1 & 0 & 0 \\ 0 & -1 & 0 \\ 0 & 0 & -1 \end{bmatrix}$$

a multiplication table of point group $2/m\text{-}C_{2h}$ and matrices of its symmetry
operations using the Cartesian coordinates of the radius vector $r = xi + yj + zk$
as a basis. For a quantum mechanic description of the central monatomic entity
with such a site symmetry then a matrix representation of dimensions 12×12 can
be constructed. When we further take as a basis Cartesian coordinates of surrounding
entities, we can assume that this matrix is of exceedingly great dimension. However,
for an adequate quantum mechanical description of such systems, representations of
far smaller dimensions are sufficient. The matrix representation of point group S
can be expressed in the form

$$\Gamma' = \left[\begin{array}{c|c} \Gamma_1(g) & Q(g) \\ \hline 0 & \Gamma_2(g) \end{array} \right] \tag{3.4.8}$$

Matrices $\Gamma_1(g)$, $\Gamma_2(g)$ and $Q(g)$ show dimension $n_1 \times n_1$, $n_2 \times n_2$ and $n_1 \times n_2$.
If the representation is unitary[1], then one can prove that $Q(g) = 0^2$ and formula
(3.4.8) may be expressed as a direct sum:

$$\Gamma'(g) = \Gamma_1(g) + \Gamma_2(g)$$

---

[1] In general: Matrix A is unitary, if the relation holds: $AA^* = I$, where A^* is a complex
conjugated matrix to A, and I is an identity matrix.
[2] See e.g. [25].

Matrix representations Γ_1 and Γ_2, being conditioned by relation $\Gamma_2 = X\Gamma_1 X^{-1}$, where X is an arbitrary matrix representation of **S** are called *conjugate*. For two arbitrary elements $\mathbf{g}_1$ and $\mathbf{g}_2$ of group **S** then one can write:

$$\Gamma_2(\mathbf{g}_1\mathbf{g}_2) = X\Gamma_1(\mathbf{g}_1\mathbf{g}_2)\, X^{-1} = X\Gamma_1(\mathbf{g})\, \Gamma_1(\mathbf{g}_2)\, X^{-1} =$$
$$= X\Gamma_1(\mathbf{g}_1)\, X\Gamma_1(\mathbf{g}_2)\, X^{-1} = X\Gamma_1(\mathbf{g}_1)\, X^{-1}\, X\Gamma_1(\mathbf{g}_2)\, X^{-1} = \Gamma_2(\mathbf{g}_1)\, \Gamma_2(\mathbf{g}_2)$$

Thus matrix Γ_2 is also a representation of group **S**. Similarly it can be proved:

(a) If matrix representation Γ_1 is conjugate to representation Γ_2, then also representation Γ_2 is conjugate to Γ_1.
(b) If matrix representation Γ_1, Γ_2, Γ_3, ... , Γ_1 are conjugate to representation Γ, then all these representations are mutually conjugate.

It follows from these properties of matrix representations that the set of all representations of point group **S** is divided into subsets of mutually conjugate representations. One can also prove that the representations of these classes of conjugate representations are unitary. For our purposes it appears important that representations in one and the same class are of the same dimensions. The matrix representation (3.4.8) then obtains the *block-diagonal form*:

$$\Gamma' = \begin{bmatrix} \Gamma_1(\mathbf{g}) & 0 & 0 & 0 & 0 \\ 0 & \Gamma_2(\mathbf{g}) & 0 & 0 & 0 \\ 0 & 0 & \Gamma_3(\mathbf{g}) & 0 & 0 \\ \vdots & & & & \vdots \\ 0 & 0 & \cdots\cdots\cdots & & \Gamma_l(\mathbf{g}) \end{bmatrix} \qquad (3.4.9)$$

or expressed using the direct sum:

$$\Gamma'(\mathbf{g}) = \Gamma_1(\mathbf{g}) + \Gamma_2(\mathbf{g}) + \Gamma_3(\mathbf{g}) + \ldots + \Gamma_1(\mathbf{g})$$

If there exists vector subspace M_1 of space M, the vectors of which are performed by representation Γ_1 into the same subspace M_1, then M_1 is *invariant* with respect to representation Γ_1. The representation Γ of group **S** is *reducible*, when in space M there exists at least one subspace M_1 being invariant with respect to all matrices $\Gamma_\mathbf{S}$. Thus, for example, the threedimensional vector space of point group $32\text{-}D_3$ (Fig. 3.4.1)

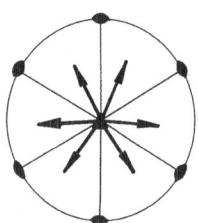

Fig. 3.4.1. Invariant subspaces of point group $32\text{-}D_3$. A stereoscopic projection. The vectors in plane are included in two dimensional subspace.

contains a onedimensional subspace on a threefold axis and a twodimensional subspace in a plane containing twofold axis, both being invariant with respect to all operations of the symmetry group. Subspace M_1 is *orthogonal* to subspace M_2 when for two arbitrary vectors $r_1 \in M_1$ and $r_2 \in M_2$ their scalar product is equal zero ($r_1 \cdot r_2 = 0$). Evidence shows that subspace M_2, orthogonal to subspace M_1 is also invariant with respect to the same representations. In space M_1 there must exist representation Γ_1 for which it holds

$$\Gamma_S r_1 = \Gamma_1(g)\, r_1$$

where $r_1 \in M_1$. When representation Γ_S is unitary, then representation Γ_1 is also unitary. Similarly in subspace M_2 representation Γ_2 can be determined applying relation

$$\Gamma_S (r_2) = \Gamma_2(g)\, r_2$$

where $r_2 \in M_2$. For each vector $r \in M$ it can always be written $r = r_1 + r_2$, or $\Gamma_S \bar{r} = \Gamma_1(g)\, r_1 + \Gamma_2(g)\, r_2$. We can say that reducible representations Γ_1 and Γ_2 are of smaller dimensions than Γ_S. If some of representations Γ_1, Γ_2 is reducible, it can be split into further representations. By this procedure we would be able to get to representations of the smallest dimensions, which cannot be split further. They are called *irreducible representations*. The vector space M will then contain subspaces $M_1, M_2, M_3, \ldots$ corresponding to irreducible representations $\Gamma_{i_1}, \Gamma_{i_2}, \Gamma_{i_3}, \ldots$. Each reducible unitary representation thus decomposes to unitary irreducible representations, this can be expressed by means of the direct sum:

$$\Gamma_S = \sum_{i=1}^{l} m_i \Gamma_i \tag{3.4.10}$$

where m_i is the number of equal irreducible representations Γ_i occurring in the matrix form (3.4.9). *The number of irreducible representations of point group* S *is equal to the number of its operation classes.* When n_i is the dimension of irreducible representation Γ_i and h is the order of point group $\check{S}$, then the relation holds

$$h = \sum_{i=1}^{l} n_i^2$$

Table 3.4.2. Symbols of irreducible representations suggested by Mulliken

Symmetry	Rotation C_n	$C_2 \perp C_n$ or $C_2\, \sigma_v(\sigma_d)$	$\sigma \perp C_n$ σ_h	Inversion center i
Symmetry	A	Subscript 1	Prime (')	g
Antisymmetry	B	Subscript 2	Doubl prime ('')	u

One dimension: A or B
Two dimensions: E
Three dimensions: T (or F)

The notation of irreducible representations according to Mulliken is shown in Table 3.4.2. In [4] are listed the irreducible representations of all point groups containing site symmetry of lattice complexes. The sum of diagonal elements of representations (trace) gives the *character of representation*, $\chi = \sum\limits_{i=1}^{n_i} \Gamma_{ii}$. The characters of irreducible representations are tabulated in [26] and [27]. The reducible representation then can be reduced using formula (3.4.10):

$$m_i = \frac{1}{h} \sum_{j=1}^{h} \chi_j \chi_j^*$$

where χ_j means the character of representation Γ_j. χ_j^* is the complex conjugate to value χ_j.

3.5 Occurence of Central Monatomic Entities in Sites of Weissenberg Lattice Complexes

The adaptability of central monatomic entities showing a certain symmetry of a crystallographic site depends on its electronic properties. This intrinsic property of central entity, for example, at the substitution in mixed crystals of one entity by another, can cause a lowering of site symmetry as well as significant distortions of the host structure [22]. Study of the probabilities of such a substitution of central monatomic entities for some structure types of mixed crystals, [23] showed a dependence of the distribution of foreign entities in crystallographic sites on the structure geometry. The distribution of foreign entities in crystallographic sites depends on the type of structure units of which the host structure consists.

Crystallographic sites of central monatomic entities may be classified based on the conception of lattice complexes. From 402 lattice complexes the Weissenberg lattice complexes have special properties already discussed. The site symmetry of such lattice complexes is described by the main group of some space group (Table 3.1.2.). It is this space group in which the lattice complex under consideration is realized. Further on let us limit our attention to the standard representation of the Weissenberg lattice complex. In this case each crystallographic site shows its highest possible site symmetry for the given lattice complex. It is the characteristic space group of this Weissenberg lattice complex. In Table 3.5.1 Weissenberg lattice complexes are summed up in their standard representations. Using the positional parameters of the central monatomic entity, the space group of the structure can then be tested to see whether it occurs in some of these lattice complexes. When a central monatomic entity occupies a site of the Weissenberg lattice complex in its standard representation, it does not have lower site symmetry than any other central entity occupying a site of another lattice complex obtained by splitting from the said Weissenberg lattice complex.

Interatomic vectors with a common origin in the site of a central monatomic entity, having symmetrically grouped their terminal entities by symmetry operations of its site symmetry, are equidistant. Their terminal points are lying on a spherical surface with its centre in the central entity site. The N-pointer thus represents a

Table 3.5.1. Weissenberg lattice complexes in standard representations [3]

Crystal system	Lattice complex[a]	Site symmetry	Multi-plicity	Symbol[b]
Triclinic	$P\bar{1}(a)$	$\bar{1}\text{-}C_i$	1	P
Monoclinic	$P2_1/m(e)$	$m\text{-}C_s$	2	$0\frac{1}{4}0\,2_1P_bACI1xz$
	$C2/c(e)$	$2\text{-}C_2$	4	$00\frac{1}{4}\bar{1}C_cF1y$
	$P2/c(e)$	$2\text{-}C_2$	2	$\frac{1}{4}0\frac{1}{4}cP_cA1y$
	$C2/m(a)$	$2/m\text{-}C_{2h}$	2	C
	$P2/m(a)$	$2/m\text{-}C_{2h}$	1	P
Orthorhombic	$P2_12_12_1(a)$	$1\text{-}C_1$	4	$2_12_1.FA_aB_aC_aI_aI_bI_cxyz$
	$Pnma(c)$	$m\text{-}C_s$	4	$0\frac{1}{4}0\bar{1}.2_1B_bA_aFI_a1xz$
	$Pbcm(d)$	$m\text{-}C_s$	4	$00\frac{1}{4}2.\bar{1}P_{bc}A_bC_cF1xy$
	$Imma(e)$	$mm2\text{-}C_{2v}$	4	$0\frac{1}{4}0.2.B_bA_a1z$
	$Cmma(g)$	$mm2\text{-}C_{2v}$	4	$0\frac{1}{4}0\,2..P_{ab}F1z$
	$Cmcm(c)$	$mm2\text{-}C_{2v}$	4	$00\frac{1}{4}2_1..C_cF1y$
	$Pmmn(a)$	$mm2\text{-}C_{2v}$	2	$2_1..CI1z$
	$Pmma(a)$	$mm2\text{-}C_{2v}$	2	P_a
	$Fddd(c)$	$\bar{1}\text{-}C_i$	16	T
	$Fddd(a)$	$222\text{-}D_2$	8	D
	$Fmmm(a)$	$mmm\text{-}D_{2h}$	4	F
	$Immm(a)$	$mmm\text{-}D_{2h}$	2	I
	$Cmmm(a)$	$mmm\text{-}D_{2h}$	2	C
	$Pmmm(a)$	$mmm\text{-}D_{2h}$	1	P
Rhombohedral	$P4_3(a)$	$1\text{-}C_1$	4	$4_3..P_{cc}{}^vDI_c1xy[z]$
	$I4_1/acd(e)$	$2\text{-}C_2$	16	$0\frac{1}{4}\frac{3}{8}\bar{4}..I_2P_{c2}1x$
	$I4_122(f)$	$2\text{-}C_2$	8	$..22^vTC_{cc}1x$
	$I\bar{4}2d(d)$	$2\text{-}C_2$	8	$\bar{4}..^vTF_c1x$
	$P4_32_12(a)$	$2\text{-}C_2$	4	I
	$P4_322(c)$	$2\text{-}C_2$	4	$00\frac{5}{8}4_3..P_{cc}{}^vD1xx$
	$P4_322(a)$	$2\text{-}C_2$	4	$00\frac{1}{4}4_3..P_{cc}I_c1x$
	$P4/nmm(c)$	$4mm\text{-}C_{4v}$	2	$0\frac{1}{2}0..2CI1z$
	$I4_1/amd(c)$	$2/m\text{-}C_{2v}$	8	^vT
	$I4_1/amd(a)$	$\bar{4}2m\text{-}D_{2d}$	4	^vD
	$I4/mmm(a)$	$4/mmm\text{-}D_{4h}$	2	I
	$P4/mmm(a)$	$4/mmm\text{-}D_{4h}$	1	P

Crystal system	Lattice complex[a]	Site symmetry	Multiplicity	Symbol[b]
Hexagonal	$P6_1(a)$	$1\text{-}C_1$	6	$3_1 2_1 .. P_{cc} E_c^{+} Q_c 1xy[z]$
	$P3_2(a)$	$1\text{-}C_2$	3	$3_2 .. P_c R^{+} Q1xy[z]$
	$P6_1 22(b)$	$2\text{-}C_2$	6	$0\,0\,\frac{1}{2}\frac{1}{2}\,3_1 2 . P_{cc} E_c^{+} Q_c 1x\bar{x}$
	$P6_1 22(a)$	$2\text{-}C_2$	6	$3_1 .2P_{cc}^{+} Q_c 1x$
	$P3_2 21(a)$	$2\text{-}C_2$	3	$0\,0\,\frac{2}{3}\,3_2 .. P_c R^{+} Q1x$
	$P3_2 12(a)$	$2\text{-}C_2$	3	$0\,0\,\frac{2}{3}\,3_2 .. P_c^{+} Q1x\bar{x}$
	$P\bar{3}m1(d)$	$3m\text{-}C_{3v}$	2	$.2.GEl\,z$
	$R\bar{3}m(e)$	$2/m\text{-}C_{2h}$	9	M
	$P6_3 22(c)$	$222\text{-}D_2$	3	^{+}Q
	$P6/mmm(f)$	$mmm\text{-}D_{2h}$	3	N
	$R\bar{3}m(a)$	$\bar{3}m\text{-}D_{3d}$	3	P
	$P6_3/mmc(c)$	$\bar{6}2m\text{-}D_{3h}$	2	E
	$P6/mmm(c)$	$\bar{6}2m\text{-}D_{3h}$	2	G
	$P6/mmm(a)$	$6/mmm\text{-}D_{6h}$	1	P
Cubic	$Ia\bar{3}(d)$	$2\text{-}C_2$	24	$.3.J_2 S^{*}V^{*}1x$
	$I\bar{4}3d(c)$	$3\text{-}C_3^2$	16	$4..I_2 Y^{**}1xxx$
	$I2_1 3(b)$	$2\text{-}C_2$	12	$2_1 3.SV1z$
	$I2_1 3(a)$	$3\text{-}C_3$	8	$2_1 2_1 .. P_2 Y^{*}1xxx$
	$P2_1 3(a)$	$3\text{-}C_3$	4	$2_1 2_1 ..FY1xxx$
	$Ia\bar{3}d(c)$	$222\text{-}D_2$	24	V^{*}
	$Ia3d(d)$	$\bar{4}\text{-}S_4$	24	S^{*}
	$Ia3d(b)$	$32\text{-}D_3$	16	Y^{**}
	$Fd\bar{3}m(c)$	$\bar{3}m\text{-}D_{3d}$	16	T
	$I4_1 32(c)$	$222\text{-}D_2$	12	^{+}V
	$I\bar{4}3d(a)$	$\bar{4}\text{-}S_4$	12	S
	$Im\bar{3}m(d)$	$\bar{4}2m\text{-}D_{2d}$	12	W^{*}
	$I4_1 32(a)$	$32\text{-}D_3$	8	$^{+}Y^{*}$
	$Fd\bar{3}m(a)$	$\bar{4}3m\text{-}T_d$	8	D
	$Im\bar{3}m(b)$	$4/mmm\text{-}D_{4h}$	6	J^{*}
	$Pm\bar{3}n(c)$	$\bar{4}2m\text{-}D_{2d}$	6	W
	$P4_3 32(a)$	$32\text{-}D_3$	16	^{+}Y
	$Fm\bar{3}m(a)$	$m\bar{3}m\text{-}O_h$	4	F
	$Pm\bar{3}m(c)$	$4/mmm\text{-}D_{4h}$	3	J
	$Im\bar{3}m(a)$	$m\bar{3}m\text{-}O_h$	2	I
	$Pm\bar{3}m(a)$	$m\bar{3}m\text{-}O_h$	1	P

[a] Wickoff letters are in parantheses

[b] Symbols used are taken from *International Tables for Crystallography* [18], Part II., pp. 823—843.

set of terminal points of N equidistant interatomic vectors. So far their site symmetries are $1\text{-}C_1$ (they occupy a general position of the point group); its order gives the number of crystallographically equidistant interatomic vectors. From this point of view, within each group of lattice complexes in Table 3.5.1 the sites of the Weissenberg lattice complexes in their standard representation are at the same time the sites with the maximum possible numbers of crystallographically equidistant interatomic vectors with their common origin in the site of central entity.

Based on the list of characteristic space groups and Wickoff positions in [3] program WEISSTAT was designed to compute the occurrence of monatomic entities in the positions of Weissenberg lattice complexes in their standard representations. In order to minimize the influence of extrinsic factors from these kinds of statistics, the computation was done only for those sets of crystal structures which have shown a significant averaging of extrinsic factors on the dispersion of interatomic vector lengths (Table 2.1.1). The structural parameters stored in file DAT of the *Cambridge Crystallographic Data Package* were used as input data. The computation results are listed in Table 3.5.2.

Table 3.5.2. Occurrence of the central monatomic entities of coordination compounds in the sites of Weissenberg lattice complexes in standard representations. Frequencies and site symmetries are given in square brackets. Weissenberg lattice complexes are ordered by increasing site symmetry. Lattice complexes with degenerate irreducible representation of site symmetry group are enclosed in the frames

Central monatomic entity	Frequencies
Ti^{3+}	$P2_12_12_1(a)$ [5; $1\text{-}C_1$]
Ti^{4+}	$P2_12_12_1(a)$ [5; $1\text{-}C_1$]
Cr^{2+}	$P2_12_12_1(a)$ [3; $1\text{-}C_1$]
Cr^{3+}	$P2_12_12_1(a)$ [15; $1\text{-}C_1$] $P\bar{1}(a)$ [2; $\bar{1}\text{-}C_i$]
Mn	$P2_12_12_1(a)$ [4; $1\text{-}C_1$] $P\bar{1}(a)$ [1; $\bar{1}\text{-}C_i$]
Mn^{1+}	$P2_12_12_1(a)$ [10; $1\text{-}C_1$] Pbcm (d) [3; $m\text{-}C_s$]
Mn^{2+}	$P2_12_12_1(a)$ [17; $1\text{-}C_1$] $P\bar{1}(a)$ [14; $\bar{1}\text{-}C_i$] $C2/c(e)$ [6; $2\text{-}C_2$] Cmma (g) [1; $mm2\text{-}C_{2v}$]
Fe	$P2_12_12_1(e)$ [29; $1\text{-}C_1$] $P2/c(e)$ [1; $2\text{-}C_2$] $C2/c(e)$ [3; $2\text{-}C_2$] Pbcm(d) [17; $m\text{-}C_s$] Pmma(e) [1; $mm2\text{-}C_{2v}$]
Fe^{1+}	$P2_12_12_1(a)$ [6; $1\text{-}C_1$] Pbcm(d) [2; $m\text{-}C_s$] Pmma(e) [2; $mm2\text{-}C_{2v}$]
Fe^{2+}	$P2_12_12_1(a)$ [28; $1\text{-}C_1$] $P\bar{1}(a)$ [11; $\bar{1}\text{-}C_i$] $P2/c(e)$ [2; $2\text{-}C_2$] $C2/c(e)$ [10; $2\text{-}C_2$] Pbcm(d) [7; $m\text{-}C_s$] Cmma(g) [1; $mm2\text{-}C_{2v}$]
Co	$P2_12_12_1(a)$ [7; $1\text{-}C_1$] Pbcm(d) [3; $m\text{-}C_s$] Pmma(e) [2; $mm2\text{-}C_{2v}$]
Co^{1+}	$P2_12_12_1(a)$ [2; $1\text{-}C_1$] $P2_1/m(e)$ [2; $m\text{-}C_s$]
Co^{2+}	$P2_12_12_1(a)$ [19; $1\text{-}C_1$] $P\bar{1}(a)$ [16; $\bar{1}\text{-}C_i$] $C2/c(e)$ [9; $2\text{-}C_2$] Pbcm(d) [3; $m\text{-}C_s$] $C2/m(a)$ [3; $2/m\text{-}C_{2h}$] Cmma(g) [1; $mm2\text{-}C_{2v}$]
Co^{3+}	$P2_12_12_1(a)$ [92; $1\text{-}C_1$] $P6_1(a)$ [3; $1\text{-}C_1$] Pbcm(d) [1; $m\text{-}C_s$] Pnma(c) [9; $m\text{-}C_s$] Cmcm(c) [1, $mm2\text{-}C_{2v}$]
Ni^{1+}	—
Ni^{2+}	$P2_12_12_1(a)$ [49; $1\text{-}C_1$] $P6_1(a)$ [1; $1\text{-}C_1$] $P\bar{1}(a)$ [48; $\bar{1}\text{-}C_i$] $P2/c(e)$ [2; $2\text{-}C_2$] $C2/c(e)$ [32; $2\text{-}C_2$] $P4_32_12$ [1; $2\text{-}C_2$] Pbcm(d) [1; $m\text{-}C_s$] Pnma(c) [6; $m\text{-}C_s$] $I2_13(b)$ [1; $3\text{-}C_3$] $C2/m(a)$ [7; $2/m\text{-}C_{2h}$] Cmcm(c) [3; $mm2\text{-}C_{2v}$]
Cu^{1+}	$P2_12_12_1(a)$ [11; $1\text{-}C_1$] $C2/c(e)$ [6; $2\text{-}C_2$] $P2_1/m(e)$ [1; $1\text{-}C_1$] Pnma(e) [1; $m\text{-}C_s$]
Cu^{2+}	$P2_12_12_1(a)$ [84; $1\text{-}C_1$] $P\bar{1}(a)$ [67; $\bar{1}\text{-}C_i$] $P2/c(e)$ [2; $2\text{-}C_2$] $C/2c(e)$ [24; $2\text{-}C_2$] Pnma(c) [11; $m\text{-}C_s$] $P2_13c(c)$ [2; $3\text{-}C_3$] $C2/m(a)$ [5; $2/m\text{-}C_{2h}$] Cmcm(c) [1; $mm2\text{-}C_{2v}$]
Zn^{2+}	$P2_12_12_1(a)$ [19; $1\text{-}C_1$] $P\bar{1}(a)$ [5; $\bar{1}\text{-}C_i$] $C2/c(e)$ [9; $2\text{-}C_2$] Pnma(c) [2; $m\text{-}C_s$] $C2/m(a)$ [1; $2/m\text{-}C_{2h}$]
Nb^{5+}	$P2_12_12_1(a)$ [19; $1\text{-}C_1$] $C2/m(a)$ [2; $2/m\text{-}C_{2h}$]

Central monatomic entity	Frequencies
Mo^{1+}	$Pbcm(d)$ $[2; m-C_s]$
Mo^{2+}	$P2_12_12_1(a)$ $[16; 1-C_1]$
Mo^{3+}	$P2_12_12_1(a)$ $[1; 1-C_1]$
Mo^{4+}	$P2_12_12_1(a)$ $[3; 1-C_1]$ $P\bar{1}(a)$ $[1; \bar{1}-C_i]$
Mo^{5+}	$P2_12_12_1(a)$ $[11; 1-C_1]$
Mo^{6+}	$P2_12_12_1(a)$ $[13; 1-C_1]$
Ru	$P2_12_12_1(a)$ $[5; 1-C_1]$
Ru^{2+}	$P2_12_12_1(a)$ $[5; 1-C_1]$ $P\bar{1}(a)$ $[1; \bar{1}-C_i]$
Rh	$P2_12_12_1(a)$ $[3; 1-C_1]$
Rh^{1+}	$P2_12_12_1(a)$ $[11; 1-C_1]$ $P4_3(a)$ $[2; 1-C_1]$
Pd^{1+}	$P2_12_12_1(a)$ $[5; 1-C_1]$
Pd^{2+}	$P2_12_12_1(a)$ $[21; 1-C_1]$ $P\bar{1}(a)$ $[22; \bar{1}-C_i]$ $P4_32_12(a)$ $[1; 2-C_2]$ $C2/m(a)$ $[1; 2/m-C_{2h}]$
Ag^{1+}	$P2_12_12_1(a)$ $[44; 1-C_1]$ $\boxed{Pm3m(a)\,[1;\, m\bar{3}m - O_h]}$
Cd^{2+}	$P2_12_12_1(a)$ $[19; 1-C_1]$ $P\bar{1}(a)$ $[5; \bar{1}-C_i]$ $P4_32_12(a)$ $[1; 2-C_2]$ $C2/m(a)$ $[1; 2/m-C_{2h}]$ $\boxed{R\bar{3}m(a)\,[1;\, \bar{3}m - D_{3d}]}$
Ce^{3+}	$P2_12_12_1(a)$ $[1; 1-C_1]$
Ce^{4+}	$P2_12_12_1(a)$ $[5; 1-C_1]$
Nd^{3+}	$P2_12_12_1(a)$ $[8; 1-C_1]$ $C2/m(a)$ $[1; 2/m-C_{2h}]$
Er^{3+}	$P2_12_12_1(a)$ $[13; 1-C_1]$ $P\bar{1}(a)$ $[1; \bar{1}-C_i]$
Ta^{5+}	$P2_12_12_1(a)$ $[2; 1-C_1]$
Re^{3+}	$P2_12_12_1(a)$ $[2; 1-C_1]$
Os	$P2_12_12_1(a)$ $[5; 1-C_1]$
Ir^{3+}	$P2_12_12_1(a)$ $[5; 1-C_1]$
Pt	$P2_12_12_1(a)$ $[3; 1-C_1]$
Pt^{2+}	$P2_12_12_1(a)$ $[26; 1-C_1]$ $P\bar{1}(a)$ $[21; \bar{1}-C_i]$ $P4_32_12(a)$ $[2; 2-C_2]$ $P4_322(c)$ $[1; 2-C_2]$ $C2/m(a)$ $[2; 2/m-C_{2v}]$
Au^{1+}	$P2_12_12_1(a)$ $[3; 1-C_1]$ $P\bar{1}(a)$ $[2; \bar{1}-C_i]$
Hg^{2+}	$P2_12_12_1(a)$ $[21; 1-C_1]$ $P3_2(a)$ $[4; 1-C_1]$ $P\bar{1}(a)$ $[8; \bar{1}-C_i]$

Statistics of the occupation of Weissenberg lattice complexes by central monatomic entities (Table 3.5.2) show an expressive preference for triclinic, monoclinic and orthorhombic lattice complexes. The other Weissenberg lattice complexes viz. tetragonal, trigonal, hexagonal and cubic are comparatively rarely occupied. The sites of those lattice complexes in standard representations have certain site symmetries. According to the Jahn-Teller theorem [28][1] a non-linear arrangement of nuclei around the central monatomic entity in an electronic degenerate state is energetically unstable. An arrangement, in which degeneration is removed, is more advantageous. A crystallographic site of the central monatomic entity, the matrix representation of which contains at least one-, two- or threedimensional irreducible

---

[1] See also [29–33].

representation (E or T), is thus energetically disadvantageous. From this point of view point groups can be selected according to the following criteria [34]:

(a) each point group is the subgroup of such a group, the matrix representation that contains at least one-, two- or threedimensional irreducible representation;
(b) a degenerate irreducible representation can be split to representations, from which at least one is nondegenerate;
(c) no subgroup is allowed if, at a gradual lowering of symmetry down to a group with non-degenerate irreducible representations, there exists at least one transition group containing non-degenerate irreducible representations only.

Table 3.5.3 presents the results of such a classification of point groups describing the symmetry of crystallographic sites. According to this classification all site-symmetry groups of triclinic, monoclinic and orthorhombic Weissenberg lattice complexes in standard representations exhibit site symmetries, the irreducible representations of which are nondegenerate (Table 3.5.1). Only two crystal structures of Cd^{2+} and Ag^{1+} compounds exhibit occupation of central monatomic entity in the site of the Weissenberg lattice complex of site symmetry with non-degenerate irreducible representations. These sites are in Table 3.5.2 closed in frames.

The structure of hexakis-(2-methylimidazole)cadmium(II)tetrafluoroborate, Cd(2-Meim)$_6$(BF$_4$)$_2$ (2-Meim = 2-methylimidazole) [35] corresponds to the first of them. This structure consists of cation units [Cd(2-Meim)$_6$]$^{2+}$ of symmetry $\bar{3}m$-D_{3d} and of anions BF$_4^-$ (Fig. 3.5.1). These units are mutually held by electrostatic forces. In spite of a high asymmetry of 2-methylimidazole ligands, the cation unit keeps its high symmetry. All entities of 2-methylimidazole ligands lie in mirror planes with Cd^{2+}. At the same time these ligands are related by a symmetry center. The Cd-N bond lengths are 2.413(3) Å (6×) being a significantly higher value than the mean value of bond lengths Cd^{2+} with imidazole nitrogen (2.361 Å) found in Cd(imidazole)$_6$(NO$_3$)$_2$

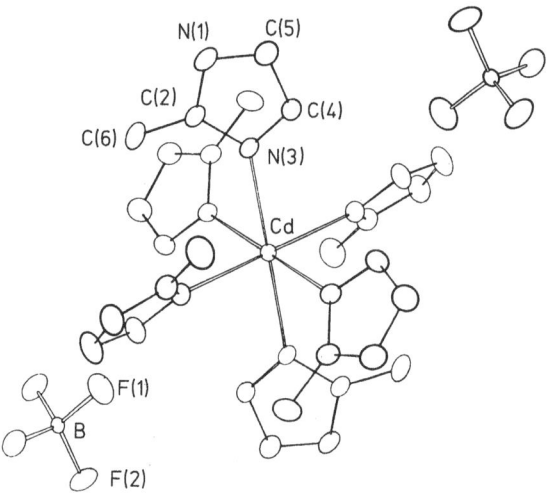

Fig. 3.5.1. Perspective view of Cd(2-Meim)$_6$(BF$_4$)$_2$ [35]. Hydrogen atoms are omitted. The trigonal axis runs through the atoms B-F(1)-Cd and through the B-F bond of the opposing BF$_4^-$ ion.

Table 3.5.3. Possible degeneracy removing of site-symmetry group irreducible representations by descent of the symmetry [34]

Site-symmetry group	Order of group	Irreducible representation	Subgroup	Order of subgroup	Stable state
$m\bar{3}m\text{-}O_h$	48	E_g	$4/mmm\text{-}D_{4h}$	16	A_{1g}, B_{1g}
			$422\text{-}D_4$	8	A_1, B_1
			$mmm\text{-}D_{2h}$	8	A_g
			$\bar{4}2m\text{-}D_{2d}$	8	A_1, B_1
			$2/m\text{-}C_{2h}$	4	A_g, B_g
			$222\text{-}D_2$	4	A
			$2\text{-}C_2$	2	A, B
			$\bar{1}\text{-}C_i$	2	A_g
			$3\text{-}C_3$	2	A', A''
		E_u	$4/mmm\text{-}D_{4h}$	16	A_{1u}, B_{1u}
			$422\text{-}D_4$	8	A_1, B_1
			$mmm\text{-}D_{2h}$	8	A_u
			$\bar{4}2m\text{-}D_{2d}$	8	A_1, B_1
			$2/m\text{-}C_{2h}$	4	A_u, B_u
			$222\text{-}D_2$	4	A
			$2\text{-}C_2$	2	A, B
			$\bar{1}\text{-}C_i$	2	A_u
			$m\text{-}C_s$	2	A', A''
		T_{1g}	$4/mmm\text{-}D_{4h}$	16	A_{2g}
			$\bar{3}m\text{-}D_{3d}$	12	A_{2g}
			$422\text{-}D_4$	8	A_2
			$32\text{-}D_3$	6	A_2
			$\bar{4}2m\text{-}D_{2d}$	8	A_2
			$mmm\text{-}D_{2h}$	8	B_{1g}, B_{2g}, B_{3g}
			$3m\text{-}C_{3v}$	6	A_2
			$2/m\text{-}C_{2h}$	4	A_g, B_g
			$222\text{-}D_2$	4	B_1, B_2, B_3
			$3\text{-}C_3$	3	A
			$mm2\text{-}C_{2v}$	4	B_1, B_2
			$2/m\text{-}C_{2h}$	4	B_g
			$2\text{-}C_2$	2	A, B
			$\bar{1}\text{-}C_i$	2	A_g
			$m\text{-}C_s$	2	A', A''
			$2\text{-}C_2$	2	B
		T_{2g}	$4/mmm\text{-}D_{4h}$	16	B_{2g}
			$\bar{3}m\text{-}D_{3d}$	12	A_{1g}
			$422\text{-}D_4$	8	B_2
			$32\text{-}D_3$	6	A_1
			$\bar{4}2m\text{-}D_{2d}$	8	B_2
			$mmm\text{-}D_{2h}$	8	B_{1g}, B_{2g}, B_{3g}
			$3m\text{-}C_3$	6	A_1
			$2/m\text{-}C_{2h}$	4	A_g, B_g
			$222\text{-}D_2$	4	B_1, B_2, B_3
			$3\text{-}C_3$	3	A
			$mm2\text{-}C_{2v}$	4	B_1, B_2
			$2/m\text{-}C_{2h}$	4	B_g
			$2\text{-}C_2$	2	A, B
			$\bar{1}\text{-}C_i$	2	A_g
			$m\text{-}C_s$	2	A', A''
			$2\text{-}C_2$	2	B

Table 3.5.3 (continued)

Site-symmetry group	Order of group	Irreducible representation	Subgroup	Order of subgroup	Stable state
		T_{1u}	4/mmm-D_{4h}	16	A_{2u}
			$\bar{3}$m-D_{3d}	12	A_{2u}
			422-D_4	8	A_2
			32-D_3	6	A_2
			$\bar{4}$2m-D_{2d}	8	B_2
			mmm-D_{2h}	8	B_{1u}, B_{2u}, B_{3u}
			3m-C_{3v}	6	A_1
			2/m-C_{2h}	4	A_u, B_u
			222-D_2	4	B_1, B_2, B_3
			3-C_3	3	A
			mm2-C_{2v}	4	B_1, B_2
			2/m-C_{2h}	4	B_u
			2-C_2	2	A, B
			$\bar{1}$-C_i	2	A_u
			m-C_s	2	A', A''
			2-C_2	2	B
		T_{2u}	4/mmm-D_{4h}	16	B_{2u}
			$\bar{3}$m-D_{3d}	12	A_{1u}
			422-D_4	8	B_2
			32-D_3	6	A_1
			$\bar{4}$2m-D_{2d}	8	B_2
			mmm-D_{2h}	8	B_{1u}, B_{2u}, B_{3u}
			3m-C_{3v}	6	A_2
			2/m-C_{2h}	4	A_u, B_u
			222-D_2	4	B_1, B_2, B_3
			3-C_3	3	A
			mm2-C_{2v}	4	B_1, B_2
			2/m-C_{2h}	4	B_u
			2-C_2	2	A, B
			$\bar{1}$-C_i	2	A_u
			m-C_s	2	A', A''
			2-C_2	2	B
$\bar{4}$3m-T_d	24	E	$\bar{4}$2m-D_{2d}	8	A_1, B_1
			222-D_2	4	A
			m-C_s	2	A', A''
		T_1	$\bar{4}$2m-D_{2d}	8	A_2
			3m-C_{3v}	6	A_2
			222-D_2	4	B_1, B_2, B_3
			3-C_3	3	A
			mm2-C_{2v}	4	B_1, B_2
			2-C_2	2	B
			m-C_s	2	A', A''
		T_2	$\bar{4}$2m-D_{2d}	8	B_2
			3m-C_{3v}	6	A_1
			222-D_2	4	B_1, B_2, B_3
			3-C_3	3	A
			mm2-C_{2v}	4	B_1, B_2
			2-C_2	2	B
			m-C_s	2	A', A''
m$\bar{3}$-T_h	24	E_g	mmm-D_{2h}	8	A_g
			222-D_2	4	A

Site-symmetry group	Order of group	Irreducible representation	Subgroup	Order of subgroup	Stable state
		E_u	$\bar{1}$-C_i	2	A_g
			mmm-D_{2h}	8	A_u
			222-D_2	4	A
		T_g	$\bar{1}$-C_i	2	A_u
			mmm-D_{2h}	8	B_{1g}, B_{2g}, B_{3g}
			222-D_2	3	B_1, B_2, B_3
			3-C_3	3	A
		T_u	$\bar{1}$-C_i	2	A_g
			mmm-D_{2h}	8	B_{1u}, B_{2u}, B_{3u}
			222-D_2	4	B_1, B_2, B_3
			3-C_3	3	A
432-O	24	E	$\bar{1}$-C_i	2	A_u
			422-D_4	8	A_1, B_1
			222-D_2	4	A
			2-C_2	2	A, B
		T_1	422-D_4	8	A_2
			32-D_3	6	A_2
			222-D_2	4	B_1, B_2, B_3
			3-C_3	3	A
			2-C_2	2	A, B
		T_2	422-D_4	8	B_2
			32-D_3	6	A_1
			222-D_2	4	B_1, B_2, B_3
			3-C_3	3	A
			2-C_2	2	A, B
23-T	12	E	222-D_2	4	A
		T	222-D_2	4	B_1, B_2, B_3
			3-C_3	3	A
6/mmm-D_{6h}	24	E_{1g}	mmm-D_{2h}	8	B_{2g}, B_{3g}
			222-D_2	4	B_2, B_3
			mm2-C_{2v}	4	A_2, B_1, B_2
			2/m-C_{2h}	4	A_g, B_g
			2-C_2	2	A, B
			$\bar{1}$-C_i	2	A_g
			m-C_s	2	A', A''
		E_{2g}	mmm-D_{2h}	8	A_g, B_{1g}
			222-D_2	4	A, B_1
			mm2-C_{2v}	4	A_1, A_2, B_2
			2/m-C_{2h}	4	A_g, B_g
			2-C_2	2	A, B
			$\bar{1}$-C_i	2	A_g
			m-C_s	2	A', A''
		E_{1u}	mmm-D_{2h}	8	B_{2u}, B_{3u}
			222-D_2	4	A, B_1
			mm2-C_{2v}	4	A_1, B_1, B_2
			2/m-C_{2h}	4	A_u, B_u
			2-C_2	2	A, B
			$\bar{1}$-C_i	2	A_u
			m-C_s	2	A', A''
		E_{2u}	mmm-D_{2h}	8	A_u, B_{1u}
			222-D_2	4	A, B_1
			mm2-C_{2v}	4	A_1, A_2, B_1
			2/m-C_{2h}	4	A_u, B_u

Table 3.5.3 (continued)

Site-symmetry group	Order of group	Irreducible representation	Subgroup	Order of subgroup	Stable state
			$2\text{-}C_2$	2	A, B
			$\bar{1}\text{-}C_i$	2	A_u
			$m\text{-}C_s$	2	A', A''
$622\text{-}D_6$	12	E_1	$222\text{-}D_2$	4	B_2, B_3
			$2\text{-}C_2$	2	A, B
		E_2	$222\text{-}D_2$	4	A, B_1
			$2\text{-}C_2$	2	A, B
$4/mmm\text{-}D_{4h}$	16	E_g	$mmm\text{-}D_{2h}$	8	B_{2g}, B_{3g}
			$222\text{-}D_2$	4	B_2, B_3
			$mm2\text{-}C_{2v}$	4	B_1, B_2
			$2/m\text{-}C_{2h}$	4	B_g
			$2\text{-}C_2$	2	B
		E_u	$mmm\text{-}D_{2h}$	8	B_{2u}, B_{3u}
			$222\text{-}D_2$	4	B_2, B_3
			$mm2\text{-}C_{2v}$	4	B_1, B_2
			$2/m\text{-}C_{2h}$	4	B_u
			$2\text{-}C_2$	2	B
$422\text{-}D_4$	8	E	$222\text{-}D_2$	4	B_2, B_3
			$2\text{-}C_2$	2	B
$\bar{6}m2\text{-}D_{3h}$	12	E'	$mm2\text{-}C_{2v}$	4	A_1, B_2
			$2\text{-}C_2$	2	A, B
			$m\text{-}C_s$	2	A', A''
		E''	$mm2\text{-}C_{2v}$	4	A_2, B_1
			$2\text{-}C_2$	2	A, B
			$m\text{-}C_s$	2	A', A''
$\bar{3}m\text{-}D_{3d}$	12	E_g	$2/m\text{-}C_{2h}$	4	A_g, B_g
			$2\text{-}C_2$	2	A, B
			$\bar{1}\text{-}C_i$	2	A_g
			$m\text{-}C_s$	2	A', A''
		E_u	$2/m\text{-}C_{2h}$	4	A_u, B_u
			$2\text{-}C_2$	2	A, B
			$\bar{1}\text{-}C_i$	2	A_u
			$m\text{-}C_s$	2	A', A''
$32\text{-}D_3$	6	E	$2\text{-}C_2$	2	A, B
$\bar{4}2m\text{-}D_{2d}$	8	E	$222\text{-}D_2$	4	B_2, B_3
			$mm2\text{-}C_{2v}$	4	B_1, B_2
			$2\text{-}C_2$	2	B
$\bar{4}\text{-}S_4$	4	E	$2\text{-}C_2$	2	B
$6/m\text{-}C_{6h}$	12	E_{1g}	$2/m\text{-}C_{2h}$	4	B_g
			$2\text{-}C_2$	2	B
			$\bar{1}\text{-}C_i$	2	A_g
			$m\text{-}C_s$	2	A''
		E_{2g}	$2/m\text{-}C_{2h}$	4	A_g
			$2\text{-}C_2$	2	A
			$\bar{1}\text{-}C_i$	2	A_g
			$m\text{-}C_s$	2	A''
		E_{1u}	$2/m\text{-}C_{2h}$	4	B_u
			$2\text{-}C_2$	2	B
			$\bar{1}\text{-}C_i$	2	A_u
			$m\text{-}C_s$	2	A'
		E_{2u}	$2/m\text{-}C_{2h}$	4	A_u
			$2\text{-}C_2$	2	A

Site-symmetry group	Order of group	Irreducible representation	Subgroup	Order of subgroup	Stable state
			$\bar{1}\text{-}C_i$	2	A_u
			$m\text{-}C_s$	2	A''
$6mm\text{-}C_{6v}$	12	E_1	$mm2\text{-}C_{2v}$	4	B_1, B_2
			$2\text{-}C_2$	2	B
			$m\text{-}C_s$	2	A', A''
		E_2	$mm2\text{-}C_{2v}$	4	A_1, A_2
			$2\text{-}C_2$	2	A
			$m\text{-}C_s$	2	A', A''
$6\text{-}C_6$	6	E_1	$2\text{-}C_2$	2	B
		E_2	$2\text{-}C_2$	2	A
$4/m\text{-}C_{4h}$	8	E_g	$2/m\text{-}C_{2h}$	4	B_g
			$2\text{-}C_2$	2	B
		E_u	$2/m\text{-}C_{2h}$	4	B_u
			$2\text{-}C_2$	2	B
$4mm\text{-}C_{4v}$	8	E	$mm2\text{-}C_{2v}$	4	B_1, B_2
			$2\text{-}C_2$	2	B
$4\text{-}C_4$	4	E	$2\text{-}C_2$	2	A
$\bar{6}\text{-}C_{3h}$	6	E'	$m\text{-}C_s$	2	A'
		E''	$m\text{-}C_s$	2	A''
$3m\text{-}C_{3v}$	6	E	$m\text{-}C_s$	2	A', A''

[36]. This difference probably originates in steric hindrances owing to the position of the methyl group in the imidazole ligand; at the same time they prevent ligands to deviate from symmetry planes. In this way the structure unit $[Cd(2\text{-Meim})_6]^{2+}$ keeps its symmetry, stabilized also by electrostatic interactions with the nearly tetrahedral units BF_4^-. The regular octahedral coordination around Cd^{2+} is slightly distorted due to being compressed in the direction of the threefold axes. The angle between these axes of symmetry and the Cd—N bond directions is then $57.05°$ instead of the value of $54.75°$ in an ideal octahedral coordination. This distortion can be caused by close packing of the cation and anion units in the crystal.

The crystal structure of silver perchlorate dioxane complex, $AgClO_4 \cdot (diox)_3$ [37] consists of Ag^{1+} entities placed in sites of the cubic lattice complex P (Fig. 3.5.2). Each central entity is surrounded by six oxygen entities of dioxan ligands in the apices of a regular octahedron. The interatomic distance Ag—O makes 2.46 Å indicating an only negligible covalent character of this bond [37]. The bidentate dioxan ligands and $[ClO_4]^-$ units placed in the centre of the unit cell are rotationally disordered thus stabilizing the most probable site symmetry of the silver entities $(m\bar{3}m\text{-}O_h)$. This symmetry is the most advantageous with respect to the Laves symmetry principle (Chap. 1.2). Since in this structure electrostatic interactions between stable structure units dominate, a dominant influence of the vector equilibrium in the crystal may be expected, thus stabilizing the site symmetry of central monatomic entities being, with respect to electronic degeneration, unstable.

The lifting of the degenerate electronic states is thus a significant factor influencing the statistics in Table 3.5.2. Using Table 3.5.4 each site symmetry group of the

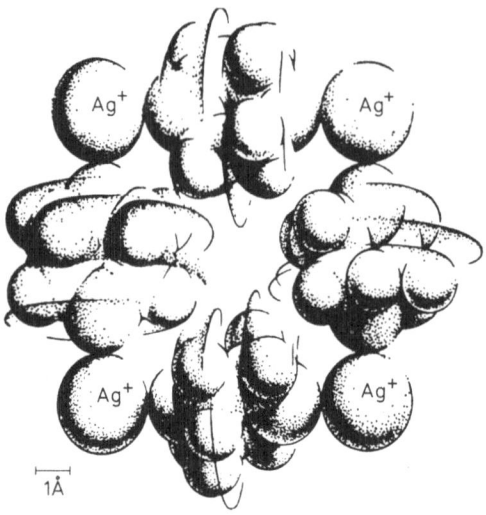

Fig. 3.5.2. A section through the crystal structure of $AgClO_4 \cdot (diox)_3$ at $x = 0$ [37]. The dioxane molecules are represented in various rotational orientations. The perchlorate ion, centered at $\left(\dfrac{1}{2}, \dfrac{1}{2}, \dfrac{1}{2}\right)$, extends to within about 1 Å of the plane $x = 0$.

Weissenberg lattice complex in standard representation with degenerate irreducible representation can be reduced to some group in Table 3.5.2 except the two site symmetries discussed. This is in agreement with the generalization deduced by Liehr [38] in the conditions of the first order perturbation theory (*minimax rule*): *Free molecule in a degenerate electronic site will descend in symmetry only so far as the nearest point group which will remove the degeneracy*. Such a descent of site symmetry can be realized by the displacing of the central monatomic entity from the position of the symmetry element, i.e. by splitting the site symmetry point, being in agreement with generalization [38] (*Principle of mathematical inherence*): *The Jahn-Teller behavior of a polyatomic group theoretic system is completely determined by its elemental subgroups, which still preserve the electronic degeneracy in question*.

Most frequently to occur are the central entities in positions of the lattice complex $P2_12_12_1(a)$. It is the only trivariant Weissenberg lattice complex (the other showing less degrees of freedom). The coordinates of the central entity position are thus not restricted by any limitations. Therefore compared with the other Weissenberg lattice complexes in standard representations, this lattice complex covers the greatest amount of possible sites. The occurrence of central entities in the sites of this lattice complex is thus most probable.

Glossary of Symbols

N	Number of equivalent points of site symmetry group
S	Intersection of symmetry elements of site symmetry group
f_S	Number of degrees of freedom

m_S	Multiplicity of the site
E	Symmetry operation of idensity
$\sigma_n, \sigma_v, \sigma_d, \sigma$	Mirror planes symmetry operations
C_n	Rotation symmetry operation
S_n	Inverse rotation symmetry operation
G	Space group
L	Lattice complex
m	Multiplicity of lattice complex
V	Volume of the unit cell
$\mathbb{H}$	Hamiltonian
r_i	Length of radius vector
$\mathbb{R}$	Symmetry operator
$\boldsymbol{e}_i$	Basic vector
M	Vector space
$\psi_i, \varphi, \vartheta$	Wave functions
δ_{ij}	Kronecker symbol
H	Matrix representation of Hamiltonion
H_{ji}	Matrix elements of Hamiltonian
$\mathbb{S}$	Point group
$\mathbf{g}$	Element of point group
Γ'	Matrix representation of point group
m_i	Number of equal irreducible representations
χ_j	Character of representation

3.6 References

1. International Tables for X-Ray Crystallography, Vol. 1, p. 25, Birmingham, Kynoch Press 1965
2. International Tables for Crystallography, Part II., p. 750, Dordrecht, Boston 1983
3. Fischer, W., Burzlaff, H., Hellner, E., Donnay, J. D. H.: Space Groups and Lattice Complexes, Natl. Bur. Stand. (US) Monogr. No. 134, Washington: National Bureau of Standards 1973
4. Salthouse, J. A., Ware, M. J.: Point Group Character Tables and Related Data, p. 65, Cambridge, Cambridge University Press 1972
5. Wiessenberg, K.: Z. Kristallogr. 62, 13 (1925)
6. Niggli, P.: Geometrische Kristallographie des Diskontinuums, Leipzig, Bornträger 1919
7. Hermann, C.: Gitterkomplexe, in: Internationale Tabellen zur Bestimmung von Kristallstrukturen, Vol. 1, Berlin, Bornträger 1935
8. Donnay, J. D. H., Hellner, E., Niggli, A.: Z. Kristallogr. 123, 255 (1966)
9. Fischer, W., Koch, E.: Z. Kristallogr. 139, 268 (1974)
10. Zimmermann, H., Burzlaff, H.: Kristallogr. 139, 252 (1974)
11. Koch, E., Fischer, W.: Acta Crystallogr., Sect. A, 41, 421 (1985)
12. International Tables for Crystallography, Part II., p. 820, Dordrecht, Boston 1983
13. Hellner, E.: Z. Anorg. Allg. Chemie 421, 49 (1976)
14. Bokij, G. B., Smirnova, N. L.: Zh. Strukt. Khim. 4, 744 (1963)
15. Hellner, E.: Z. Anorg. Allg. Chem. 421, 41 (1976)
16. Fischer, W.: Z. Kristallogr. 140, 50 (1974)
17. Loeb, A. L.: J. Solid State Chem. 1, 237 (1970)
18. International Tables for Crystallography, Part II., p. 724, Dordrecht, Boston 1983
19. International Tables for X-Ray Crystallography, Vol. 1, p. 74, Birmingham, Kinoch Press 1965

20. Nowacki, W.: Z. Kristallogr. **73**, 331 (1933)
21. Koch, E.: Z. Kristallogr. **138**, 196 (1973)
22. Reinen, D., Friebel, C.: Structure and Bonding **37**, 1 (1979)
23. Friebel, C.: Acta Crystallogr., Sect. A, **36**, 259 (1980)
24. Birkhoff, G., McLane, S.: A Survey of Modern Algebra, New York, MacMillan Co. 1960
25. Hall, L. H.: Group Theory and Symmetry in Chemistry, p. 111, New York, St. Louis, San Francisco, London, Sydney, Toronto, Mexico, Panama, McCraw-Hill Book Co. 1969
26. Wilson, E. B. Jr., Decius, J. C., Cross, P. C.: Molecular Vibrations, p. 323, New York, McCraw-Hill Book Co. 1955
27. Salthouse, J. A., Ware, M. J.: Point Group Character Tables and Related Data, p. 33, Cambridge, Cambridge University Press 1972
28. Jahn, H. A., Teller, E.: Proc. Roy. Soc **A 161**, 220 (1937)
29. Ballhausen, C. J.: Introduction to Ligand Field Theory, New York, McCraw-Hill Book Co. 1962
30. Bersuker, I. B.: Elektronnyje strojenija i svojstva koordinacionnych sojedinenij, Leningrad, Chimija 1976²
31. Ham, F. S.: Electron Paramagnetic Resonance, (ed.) Geschwind, J., New York, Plenum Press 1971
32. Robinson, G. W.: Methods of Experimental Physics, in: Molecular Physics, (ed.) Williams, D., p. 155, New York, London, Academic Press Inc. 1962
33. Birman, J. L.: Phys. Rev. **125**, 1959 (1962)
34. Pelikán, P., Breza, M.: Chem. Papers **39**, 255 (1985)
35. Reedijk, J., Verschoor, G. C.: Acta Crystallogr., Sect. B, **29**, 721 (1973)
36. Mighell, A. D., Santoro, A.: Acta Crystallogr., Sect. B, **27**, 2089 (1971)
37. Prosen, R. J., Trueblood, N.: Acta Crystallogr. **9**, 741 (1956)
38. Liehr, A. D.: J. Phys. Chem. **67**, 389 (1963)

4 Statistics of Bond Distances Between Central Monatomic Entities and Ligands

The search for the factors causing distortions of surroundings of central entities is the subject of interest especially of chemists. It is the region of interatomic vector lengths, which are less than the most probable minimum value $a_0' = 4.25$ Å, where it is already difficult to suppose a fitting of some known statistical model. The non-rigid properties of central entities in this region, let us call it the *coordination sphere* of entity M^{z+}, apparently manifest themselves also in distortions of the geometry of arrangement of monatomic ligand entities. In agreement with the commonly used terms for describing crystal structures of compounds we will call monatomic entities of ligands "ligand atoms" resp. atoms. The number of bonds formed by the central entity is the coordination number. The bond lengths between the central entity and ligand atoms will be signed M-L. The central entity with atoms coordinated to it forms the *inner coordination sphere*. For many coordination compounds in condensed matter it is, however, problematic to determine, whether the ligand atom under consideration is coordinated to the central entity or not. The intricacy of this problem is also shown by the mostly complicated form of M^{z+} histograms (pp. 27–53) in the region of interatomic vector lengths being shorter than $a_0' = 4.25$ Å. The concept of coordination and of coordination number based on the close packing of ligand atoms around the central entity [1, 2] appears here to be an insufficient approach. The *bond force* [2] or *bond valence* can be expressed (according [3]) by the relation:

$$s = s_0 \left(\frac{R}{R_0} \right)^{-N}$$

or

$$s = e^{\frac{R - R_0}{B}} \tag{4.1}$$

where s is bond valence and R is bond length M-L; s_0 is the valence of standard bond length R and N, B are constants, respectively. These coefficients in formula (4.1) are chosen in a way that will satisfy the *valence sum rule* as exactly as possible [3, 4]

$$z = \sum_i s_i \tag{4.2}$$

where s_i means the valence of ith bond around the central entity of theoretical valence z. The fitting of function (4.1) by inserting it into (4.2) for many crystal structures led to the formulation of an *empirical bond valence theory* [4]. Although this theory has no general validity, especially for central entities with typical electronic or steric effects, the convex shape of the curve *s vs. R* for the most part of *M-L* bonds may be presupposed. From this intrinsic property of the central entity the corrolary follows (*the distortion theorem*) [4]: *In any coordination sphere in which the sum of the valences at the central entity is kept constant, any deviation of the individual bond length from their average value will increase the average bond length.*

For the aim of studying the causes of lowering the site symmetry of central entities a statistical analysis of bond lengths *M-L* appears to be suitable. We will use the sample of bonds which have the same central entity M^{z+} (in certain spin state) and the same element of *L*. Estimates of variances of such samples with their central entity occupying a site of the Weissenberg lattice complex in standard representation were calculated. We will select from this set of crystal structures those compounds showing a significant deviation of at least one bond length *M-L* in an inner coordination from its expected value. According to the distortion theorem such a deviations leads to a deformation of other bond lengths *M-L* in the inner coordination sphere. For the significance testing of bond lengths deviations from the sample means were used characteristics T_{n-1} (Chap. 1.1) and $\alpha = 0.05$. Only bond lengths samples with significant size were used[1].

It may be expected that the philosophy of these examples of structures will bring information of possible causes of the distortions in coordination spheres and thus also the causes of lifting the degeneracy of electron states. Irreducible representations of polynuclear aggregates are given by their complex symmetry group and therefore also by their *non-crystallographic symmetry*. Thus an analysis of these structures from the said aspects can give a picture of the origin of the geometry distortions of the crystal structure units.

A significant factor influencing the bond lengths *M-L* is the filling of energy levels of central entity with electrons at a certain kind of acting of ligands. Spin multiplicity [5] or the spin state is a fact which has been considered in calculating the statistical characteristics.

In many cases contact of central entity with a carbon atom of cyclic ligand indicates a π-bond. For such a reason a selection of bond lengths *M-C* could not be unique. That is why a bond length of the *M-C* type were not taken into account. The occupying of energy levels of central entities in cases of coordination compounds can be reliably determined also in their condensed matter by applying the half-empirical qualitative theory of crystal or ligand field which will be discussed further on.

The examples of crystal structures, selected using described criteria applied to sets of bond lengths *M-L*, are further ordered by the decreasing number of structures and by the increasing atomic number of their central entities. The clinographic figures of structure parts correspond to the drawings made by program PLUTO [6].

Reliability of crystal structure analyses will be expressed by crystallographic R-value:

$$R = \sum_{hkl} \left\| F_o(hkl) \right| - \left| F_c(hkl) \right\| \bigg/ \sum_{hkl} \left| F_o(hkl) \right|$$

---

[1] Calculated by formula (1.1.9) using $\Delta = 0.05$ Å

4.1 Terms of Monatomic Entities of Transition Elements and Splitting of Their Energy Levels in the Crystal Field

In ions of transition elements, all orbitals (except completely occupied, the highest d ones) are either or they are entirely vacant. The metal cation in an ionic crystal or in a mononuclear complex is surrounded by its nearest neighbours which can be negative ions or dipoles, both of them being directed with their electrons against the cation. All five d orbitals in a free atom or ion have the same energy and so they are degenerate. The presence of negative charges of ligands, however, removes this degeneration.

It was Bethe [7] who used the theory of crystal field in its primary form to explain the splitting of energy levels of d orbitals (Fig. 4.1.1). He conceived that ligands on X, Y and Z axes act as negative point charges. These charges influence the d orbitals lying in directions of the axes, much more, than those between the axes and those forming with them angles of $45°$. Consequently in an octahedral field for example, the energy of orbitals e_g will be higher than that of orbitals t_{2g} (Fig. 4.1.2).

The model of point charge in Bethe's theory of crystal field has no physical justification, nevertheless, based on the conception of coulomb interactions one can obtain qualitatively right results. The term *ligand-field theory* is now usually applied to a modification of the *crystal-field theory* which semiempirically takes into account the partially covalent character of the metal-ligand bond. The crystal-field theory gives a useful account of spectral and magnetic properties of coordination compounds[1].

It is well-known that the ligand-field strength (Δ_0) depends on the electron density provided by ligands in the region where d — orbitals are spread. The strength is great when the free electron pairs of ligands are in suitably directed orbitals, but it is smaller in the case of ligands, e.g. of co-ligands where free electron pairs are not expressively directed. For the given cation the extent of splitting by ligands increases in the order:
$I^- < Br^- < CrO_4^{2-} \sim Cl^- < dsep^- \sim S^{2-} \sim dtp^- < N_3^- < F^- < dtc^- < $ urea
$\sim OH^- \sim IO_3^- \sim$ oxalate$^{2-} \sim$ malonate$^{2-} \sim O^{2-} \ll H_2O < SCN^- <$ py
$\sim NH_3 <$ en $\sim SO_3^{2-} < NO_2^- \sim$ dpy $\sim$ phen $< CH_3^- \sim C_6H_5^- < CN^- \ll$ constrained phosphites [9] i.e. the known spectrochemical order. This order is nearly independent on the type of central atom, however, the precise value of Δ_0 naturally expresses a certain dependence in this direction.

For octahedral complexes of ions with electron configuration d^4 (e.g. Mn^{3+}) essentially two electron configurations are possible (Fig. 4.1.2).

Whether ion complexes d^4, with octahedral configuration ions, are configured by mode (a) or (b) respectively, the value of Δ_0 decides. In a strong ligand field, for example of CN^- the energy decrease caused by the electron transition from orbital e_g to that of t_{2g} is greater than it is necessary for the compensation of the increase of Coulomb and exchange energy. Thus for these kind of complexes, e.g. $[Mn(CN)_6]^{3-}$, occupation according to scheme (b) takes place and only two unpaired

---

[1] The theory of the ligand field was discussed in detail in several papers, e.g. [8]

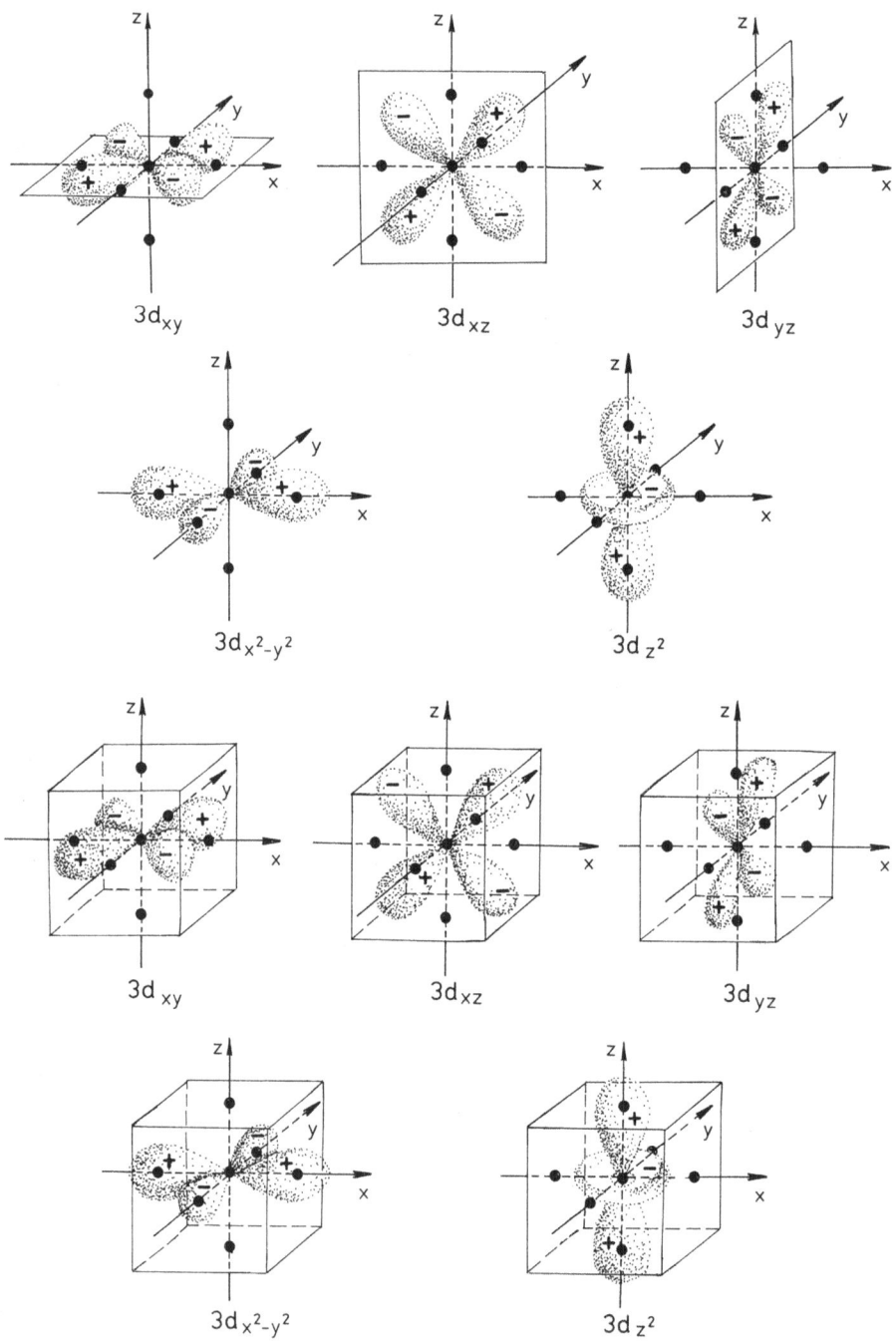

Fig. 4.1.1. Schematic representation of $3d$ orbitals

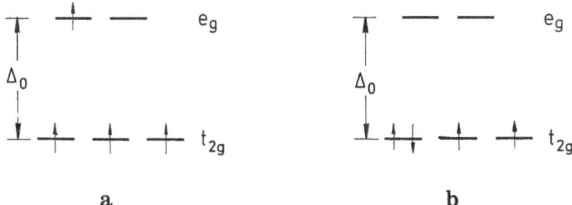

Fig. 4.1.2a, b. Two possible configurations of four d electrons in octahedral field.
(a) Coulomb repulsion is minimum and six possible electron pairs with parallel spins stabilize the ion.
(b) Coulomb repulsion is stronger, since two electrons occupy the same orbital and thus only three pairs with parallel spins can exist.

spins contribute to the magnetic moment; the respective complex is called *low-spin complex*.

On the other hand, for example, chloride anions in the function of ligands form a weaker field, Δ_0 is small, so that the occupation of orbitals takes place according to scheme (a). In this case already four unpaired spins contribute to the magnetic moment and the complex is then called *high-spin* complex. For octahedral complexes of ions with electron configuration d^5 (e.g. Fe^{3+}) that according to scheme (b) applies, in a strong ligand field, due to the transition of both electrons from orbital e_g to that of t_{2g} from two possible arrangements of electrons (a) and (b) (Fig. 4.1.3)

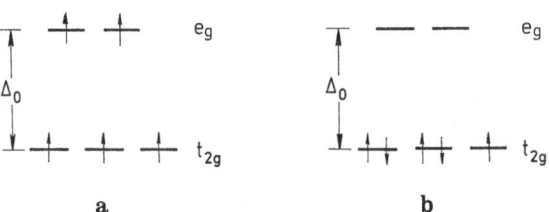

Fig. 4.1.3a, b. Two possible configurations of five d electrons in octahedral field.

scheme (b) applies. The value of magnetic moment in such a case corresponds to the presence of one unpaired electron and it is then a low-spin complex. A field formed, for example, by water molecules is, however, far too weak to realize such a pairing; the value of magnetic moment corresponds to five unpaired spins (Fig. 4.1.3(a)) and the complex shows highspin character.

High-spin and low-spin octahedral complexes are known also for ions with configuration d^6 and d^7, respectively. As to the electron configuration of ions in complexes with octahedral coordination the readiness of spin-pairing increases in the order: $d^7 < d^4 \sim d^5 < d^6$.

Symmetrical nonlinear molecules in a degenerate electron state are not stable and therefore they undergo a certain deformation in order to lower their symmetry and thus to lift up the degenerate state. This effect can be observed [10] for complexes, the central entity of which has its degenerate orbitals non uniformly occupied.

Let us demonstrate this effect on the example of an octahedral complex of ion Cu^{2+} (d^9). We shall assume that orbital $d_{x^2-y^2}$ of Cu^{2+} ion is occupied by one electron only, while in orbital d_{z^2} is a pair of electrons. In this case of Cu^{2+} octahedral complex the two ligands in the direction of axis Z are more shaded off from the nucleus charge of copper(II) atom than the remaining four ligands lying on axes X and Y. This leads generally to an increase of the bond length copper — a ligand on axis Z compared with the distance of these particles on axes X and Y, resulting in a distortion of the octahedron; the state with an unpaired electron in orbital $d_{x^2-y^2}$ becomes energetically lower. An analogous process is distorted by shortening the distance in direction of axis Z, however, it is then the state with the unpaired electron in the orbital d_{z^2} that is energetically lower. Owing to a consequence of the Jahn-Teller effect the octahedral coordination corresponding to point group O_h is changed to tetragonal bipyramidal coordination of symmetry D_{4h}. The Jahn-Teller effect also applies, for example, for octahedral low-spin complex ions of configuration d^7 (Co^{2+}, Ni^{3+}) and for high-spin complex ions of configuration d^4 (Cr^{2+}, Mn^{3+}).

The energy splitting of orbitals d in tetrahedral field is schematically shown in Fig. 4.1.4. As it is to be seen from this, the group t_{2g} covers orbitals d_{xy}, d_{yz} and d_{zx}, having a higher energy, caused by the fact that with their position they are nearer to the ligands than orbitals of groups e_g, i.e. $d_{x^2-y^2}$ and d_z^2.

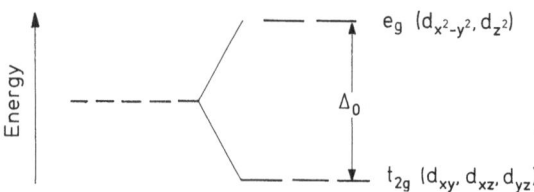

Fig. 4.1.4. Energy splitting of d orbitals in tetrahedral field.

There is an important quantitative difference between tetrahedral and octahedral compounds. Octahedral compounds possess six charges, whereas tetrahedral exhibit only four. Further the four charges are not arranged so efficiently as in the octahedral case, i.e. the metal orbitals do not joiner directly at the ligands. Dunn et al., [11] have shown that this tends to reduce the crystal field splitting, relatively, to the octahedral case by a factor of approximately 2/3.

In certain cases the number of unpaired electrons is determined by the extent of the crystal field splitting. Let us consider the octahedral case. The lower one-electron level is triply degenerate, t_{2g}. In cases d^1, d^2 and d^3 electrons occupy the t_{2g} level as the lowest energy state. However, in cases d^4–d^7 a choice is possible, as shown in Fig. 4.1.5.

A square complex can be considered as an octahedral configuration which underwent such a great tetragonal distortion most frequently in direction of axis Z, so that both ligands lying on this axis were practically shifted away from the central atom. In the field formed in this way the energy of orbitals $d_{x^2-y^2}$ reaches a higher level (Fig. 4.1.6). The energy set free at the transition of the electron from this orbital $d_{x^2-y^2}$ to orbital d_{xy} is approximately the same as in the case of octahedral field of the same strength. The tendency to form square complexes if found for ions with the configuration d^8 (Ni^{2+}, Pd^{2+}, Pt^{2+}) can here remain unoccupied since the

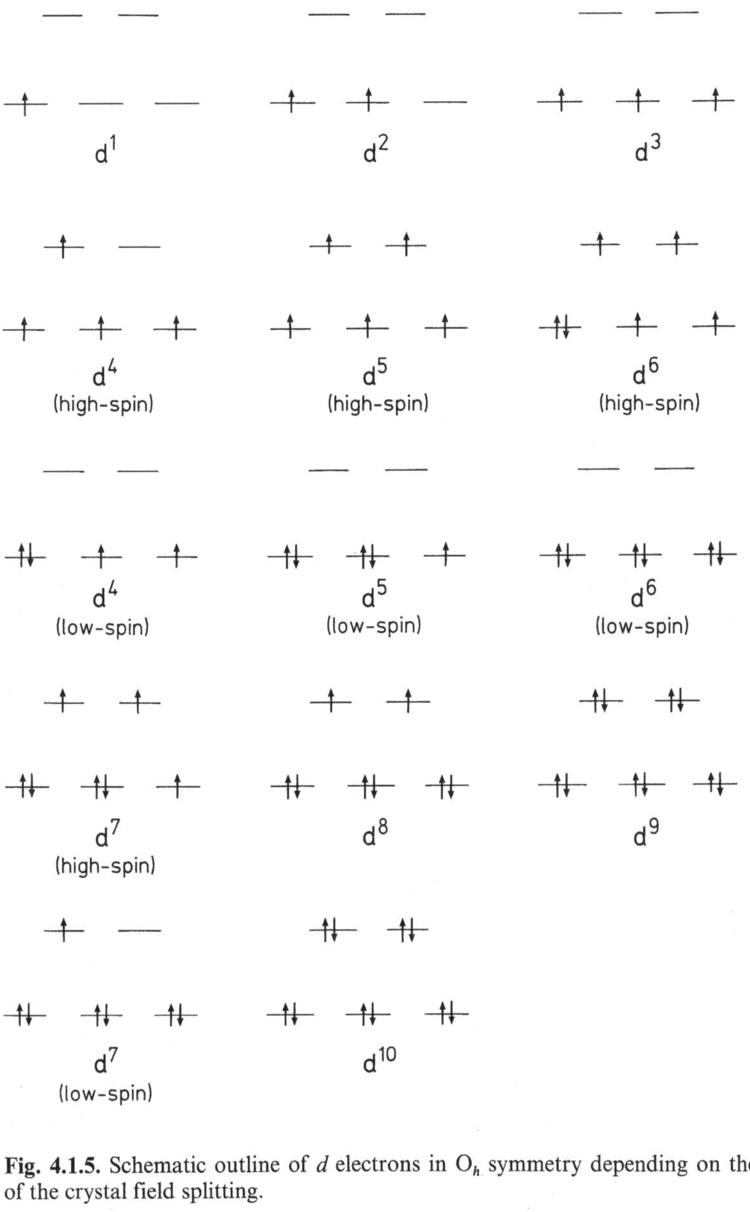

Fig. 4.1.5. Schematic outline of d electrons in O_h symmetry depending on the relative extent of the crystal field splitting.

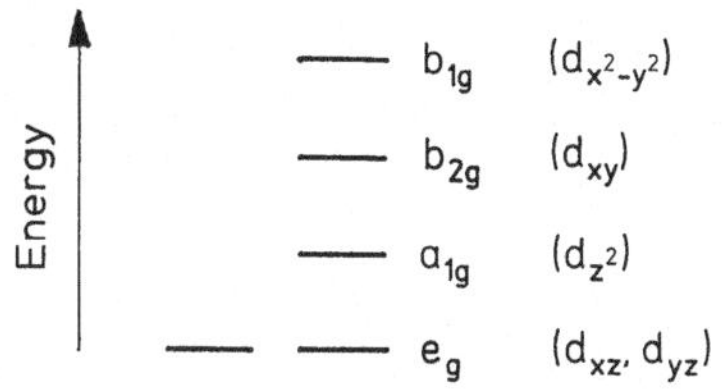

Fig. 4.1.6. The order of energies in a square complex.

energetically disadvantageous orbital is $d_{x^2-y^2}$, while a considerable amount of energy is set free by the transition of electron from this orbital to orbital d_{xy}. Such a splitting corresponds to the symmetry of point group D_{4h}.

4.2 Cu(II) Compounds

The Cu^{2+} histogram, as it is to be seen (p. 35) exhibits the expressively highest maximum with middle of class, viz., 1.975 Å and another smaller maximum with its middle 2.325 Å still belonging in the region of the atom radii sum of the central entity and of donor ligand atoms (Table 1.1). A comparatively high value of the variance estimate σ^2 of this entity (Fig. 2.2.4) apparently is given by the relatively great distance of the first maximum from the mean value of this empirical distribution. The existence of the second maximum is in agreement with the fact that in this case of central entity there exists a rather great quantity of crystal structures with bond lengths of copper — ligand atom considerably deviating from the mean value of the other bond lengths of the inner coordination sphere.

The crystal structure of diaquobis(2-hydroxy-2-methylpropionato)copper(II) [12] consists of discrete structure units [Cu(hmpr)$_2$(H$_2$O)$_2$] (Fig. 4.2.1) mutually bonded by hydrogen bonds. Entities of Cu^{2+} occupy symmetry sites $2/m$-C_{2h} of the space group C2/m in the way, that the chelate ring atoms of ligands lie in mirror planes (Fig. 4.2.2). The central entity is coordinated by four oxygen atoms of hmpr ligands in trans-arrangement with bond lengths of Cu-0(1) 1.89(2) Å (2×) and Cu-0(3) 2.01(2) Å (2×), this difference not being significant. The coordination of entity Cu^{2+} is completed by two axial bonds of oxygen atoms from two water molecules with

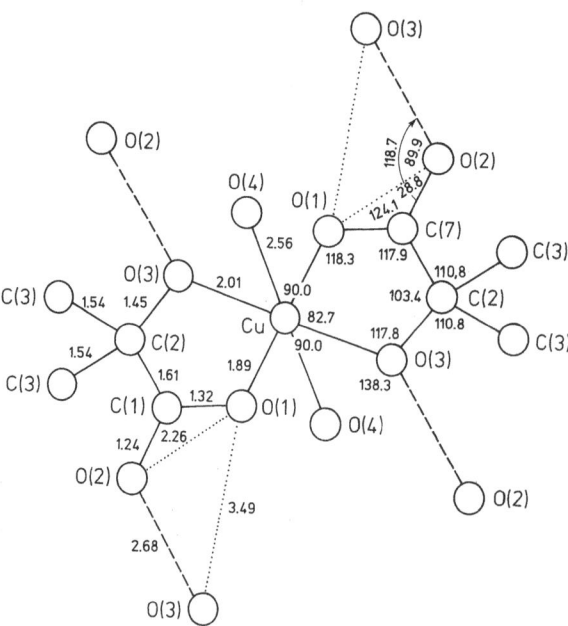

Fig. 4.2.1. Interatomic distances and angles in the structure of Cu(hmpr)$_2$(H$_2$O)$_2$ [12].

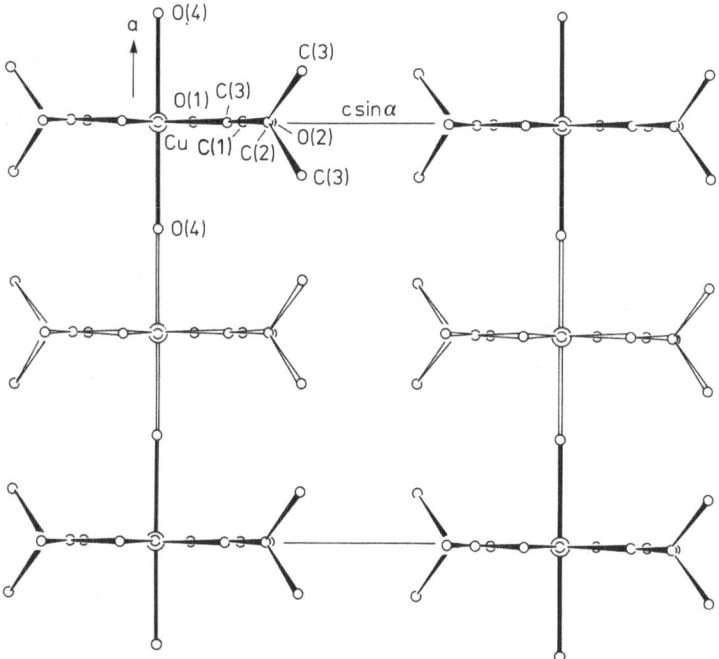

Fig. 4.2.2. Projection of crystal structure of $Cu(hmpr)_2(H_2O)_2$ [12]. Space group C2/m, monoclinic, $Z = 2$, $a = 10.25(5)$, $b = 5.80(3)$, $c = 11.24(5)$ Å and $\beta = 105.9(5)°$; $R = 0.129$.

bond lengths of Cu-O(4) 2.56(3) Å ($2 \times$). Though the coordination of Cu^{2+} may be considered tetragonal-bipyramidal of symmetry D_{4h}, no significant distortion takes place owing to site symmetry of the central entity ($2/m$-C_{2h}), since it does not contain any degenerate irreducible matrix representation. This site symmetry apparently is given by the low symmetry of hmpr ligands (C_s) as also by their coordination mode. The mutual *cis*-arrangement of chelate ligands namely, would not mean an increase of site symmetry of the central entity.

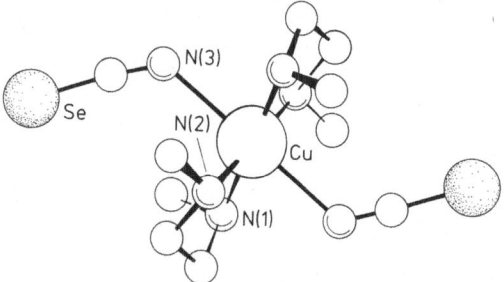

Fig. 4.2.3. Structure unit of $[Cu(Me_2en)_2(NCSe)_2]$ [13]. Space group P$\bar{1}$, triclinic, $Z = 1$, $a = 8.071(1)$, $b = 12.451(2)$, $c = 7.950(1)$ Å, $\alpha = 132.13(1)°$, $\beta = 112.94(1)°$ and $\gamma = 100.14(2)°$; $R = 0.062$.

Bis(N,N'-dimethylethylenediamine)copper(II)selenocyanate shows its central entity surrounded by N-donor ligand atoms; Me_2en and NCSe. The crystal structure of this compound [13] consists of structure units $[Cu(Me_2en)_2(NCSe)_2]$ (Fig. 4.2.3), mutually held together by Van der Waals forces. The coordination of Cu^{2+} is distorted octahedral with bond lengths of Cu-N(1) 2.05(1) Å ($2\times$), Cu-N(2) 2.07(1) Å ($2\times$) and Cu-N(3) 2.56(2) Å ($2\times$). The bond angles N(1)-Cu-N(2), N(1)-Cu-N(3) and N(2)-Cu-N(3) are 85.0(4)°, 82.1(5)° and 85.4(5)°. The selenocyanate groups exhibit a significant deviation from linearity (the bond angle N-C-Se being 176.2(2)°). Similarly the angle Cu-N(3)-C(5) is 126(1)°. These deviations from linearity are probably caused by close packing of the structure units $[Cu(Me_2en)_2$-$(NCSe)_2]$, which is also responsible for the distortion of bond angles N-Cu-N from their tetragonal values. Thus the degeneration of electron states of the central entity is removed by descending its site symmetry to $\bar{1}$-C_i. The rocked conformation of chelate ligands Me_2en and their steric effects given by the positions of methyl groups apparently also contribute to lowering the possible higher site symmetry of the central entity.

A distored octahedral coordination with chromophore CuN_4O_2 is exhibited also by entity Cu^{2+} in the crystal structure of L-histidinato-D-histidinato diaquocopper(II) tetrahydrate [14]. The structure consists of $[Cu(his)_2(H_2O)_2]$ units and of water molecules (Fig. 4.2.4). Cu^{2+} is *trans*-coordinated by two (his) chelate ligands via their N(amino) and N(imidazole) atoms. Bond lengths of Cu-N(1) and Cu-N(2) are 2.033 Å ($2\times$) and 1.996 Å ($2\times$), while the bond angle N(1)-Cu-N(2) makes 90.6° ($2\times$). The approximately square-planar coordination of Cu^{2+} is completed by two oxygen atoms from the water molecules in axial direction. The bond length of Cu-0(1) makes 2.570 Å ($2\times$). The degeneration lifting of Cu^{2+} electron states in this compound is apparently realized by lowering the site symmetry as a consequence of asymetrical his moieties. The angle between normals of the plane of imidazole

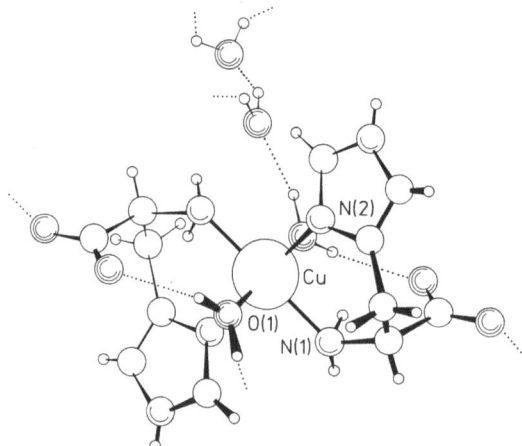

Fig. 4.2.4. Structure unit of $[Cu(his)_2(H_2O)_2]$ and schematic drawing of hydrogen bonds in the structure of $[Cu(his)_2(H_2O)_2] (H_2O)_2$ [14]. Space group P$\bar{1}$, triclinic, $Z = 1$, $a = 6.894(1)$, $b = 8.478(1)$, $c = 9.553(1)$ Å, $\alpha = 98.61(1)°$, $\beta = 116.13(1)°$ and $\gamma = 95.88(1)°$; $R = 0.076$.

ring and the plane of CuN_4 is 17°. The bond angles 0(1)-Cu-N(1) and 0(1)-Cu-N(2) are 94.2° (2×) and 90.6° (2×). The centrosymmetrical structure of unit $[Cu(his)_2(H_2O)_2]$ is also stabilized by a system of inter- and intramolecular hydrogen bonds which apparently is also responsible for a small deviation of bond angles O-Cu-N from 90.0°.

Similarly for the crystal structure of bis(N-ethylethylenediamine) copper(II) perchlorate [15] consisting of centrosymmetrical units of $[Cu(Eten)_2(ClO_4)_2]$ (Fig. 4.2.5) one can conclude that the lifting of degeneration was realized by asymmetry of chelate Eten ligands. The bond lengths of Cu-N(1) and Cu-N(2) in equatorial plane are 2.013(3) Å (2×) and 2.031(2) Å (2×), respectively, thus showing here a significant difference. In axial direction the bond lengths of Cu-O(2) make 2.594(3) Å (2×).

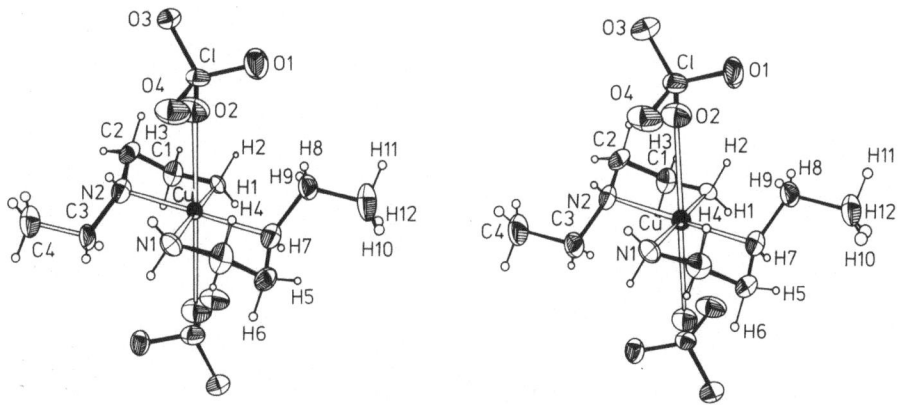

Fig. 4.2.5. Stereoscopic view of the structure unit of $[Cu(Eten)_2(ClO_4)_2]$ [15].

The chelate Eten ligand is of expressively rocked conformation, this being probably the cause of the deviation of bond angles N(1)-Cu-N(2) of both metallocycles from their tetragonal values to 85.0(1)° (2×) (Fig. 4.2.5). The structure of $[Cu(Eten)_2(ClO_4)_2]$ is stabilized by a system of hydrogen bonds in which oxygen atoms of perchlorate groups act as acceptors. These bonds and the close packing of structure units in the crystal (Fig. 4.2.6) are probably responsible for the deformation of bond angles N(1)-Cu-O(2) and N(2)-Cu-O(2) from 90° to 90.1(1) (2×) and 81.6(1)° (2×).

The degeneration lifting of electron states of the central entity as a consequence of the asymmetry of chelate-bonded bidentate ligands with Cu^{2+} is apparent also in the structure of bis(N-methylenediamine)copper(II) adipate dihydrate [16]. The structure consists of centrosymmetrical cation units $[Cu(Meen)_2(H_2O)_2]^{2+}$ and of adipate anions (Fig. 4.2.7). The anion structure units are placed between the cation units (Fig. 4.2.8) and are bonded with them by hydrogen bonds. The entity coordination is again distorted-octahedral with an elongated axial bond Cu-O(1) of the length of 2.561(2) Å (2×). The bonds of Cu-N(1) and Cu-N(2) in the equatorial plane of the coordination polyhedron around Cu^{2+} are 2.048(2) (2×) and 2.008(2) Å (2×), respectively. The small, however, significant elongation of the first bond

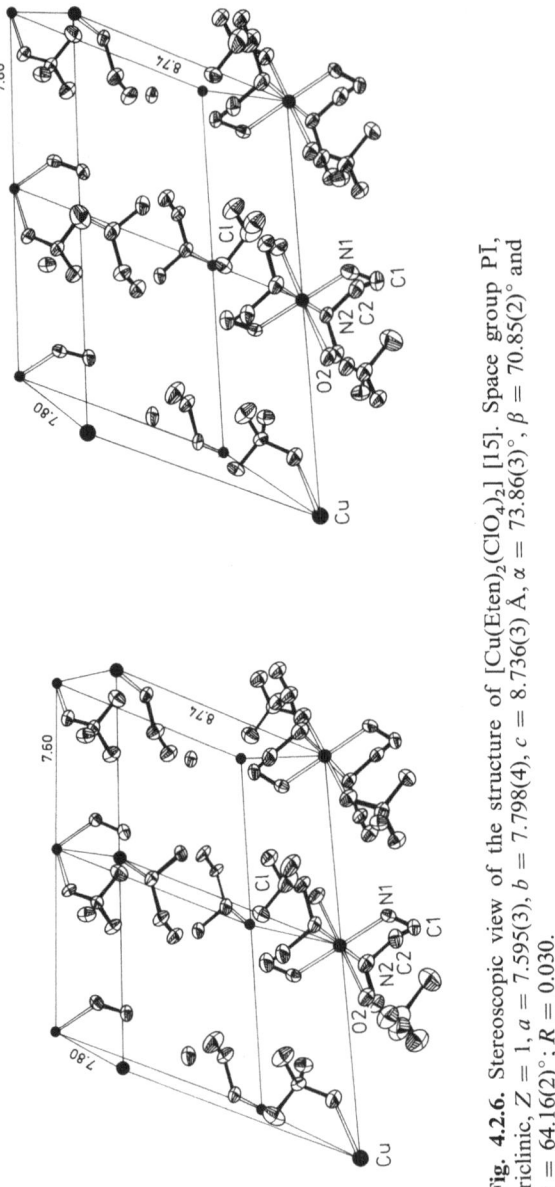

Fig. 4.2.6. Stereoscopic view of the structure of [Cu(Eten)$_2$(ClO$_4$)$_2$] [15]. Space group P$\bar{1}$, triclinic, $Z = 1$, $a = 7.595(3)$, $b = 7.798(4)$, $c = 8.736(3)$ Å, $\alpha = 73.86(3)°$, $\beta = 70.85(2)°$ and $\gamma = 64.16(2)°$; $R = 0.030$.

apparently is caused by steric influence of the methyl group bonded to N(1) atom in ligand Meen. The *gauche* conformation of the diamine ligand probably is the cause of the deviation of the bond angle N(1)-Cu-N(2) from its tetragonal value in the pentametallocycle to that of 85.4(1)° (2×). To the stabilization of the elongated bond Cu-O(1) in axial direction to the lengths of 2.561(2) Å (2×) apparently also contribute hydrogen bonds of coordinated water molecules with uncoordinated adipate anions. This circumstance is probably responsible also for the deviation

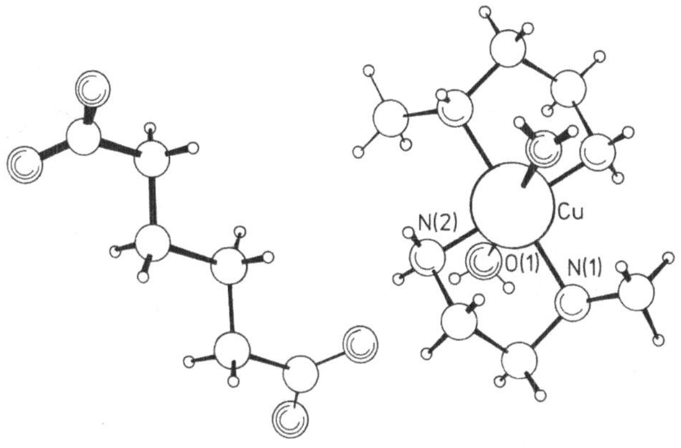

Fig. 4.2.7. The structures of the cationic unit of [Cu(Meen)$_2$(H$_2$O)^2]$^{2+}$ and of adipate anions [16].

of bond angles O(1)-Cu-N(2) and O(1)-Cu-N(2) from the tetragonal values to 88.9(1) (2×) and 94.0(1)° (2×).

In the crystal structure of dinitratobis[2-(2-diethylammonioethyl)pyridine]copper-(II)nitrate [17] the low site symmetry of the central entity is formed by monodentate or potentially bidentate DEAEPH ligand coordinated to central entity as also by bidentate nitrate groups coordinated to the Cu^{2+} entity by bonds Cu-O(1) and Cu-O(2), respectively (Fig. 4.2.9) showing quite different bond lengths. The first of them having the value 2.562(3) Å (2×), while the second 1.992(2) Å (2×). The DEAEPH moiety coordinated to Cu^{2+} by bond Cu-N(1) of the length 1.984(2) Å (2×), has its free part containing uncoordinated nitrogen atom N(2) stabilized by intermolecular hydrogen bond of this atom with the uncoordinated nitrate group. Within one [Cu(DEAEPH)$_2$(NO$_3$)$_2$] structure unit the Cu^{2+} entity is hexacoordinated by two nitrogen atoms from planar pyridin rings of DEAEPH moieties and by two oxygen atoms of nitrate groups nearly to square-planar form; furthermore, two oxygen atoms of the same nitrate groups occupy coordination sites five and six. The bond angle N(1)-Cu-O(1) is 92.29(9)° (2×). This small, however, significant deviation from 90° probably is caused by the influence of close packing of [Cu(DEAEPH)$_2$(NO$_3$)$_2$] on the rigid nitrate groups. The acute bond angle of O(1)-Cu-O(2) with its value 54.85(8)° (2×) is apparently given by stereochemical possibilities of nitrate groups. The monotonic decreasing of bond lengths 1.283(3), 1.237(3), and 1.218(3) Å for N(3)-O(2), N(3)-O(1) and N(3)-O(3), respectively, is in agreement with the fact that atom O(2) is strongly bonded, while O(1) only weakly bonded with central entity and O(3) is even non-bonded. This fact indicates a certain stress in nitrate groups and also the tendency of this ligand to lower the site symmetry of Cu^{2+}. Since rigid pyridine rings have the possibility of rotation around axis Cu-N(1), DEAEPH has rather a stabilizing effect on the complex structure.

The *gauche* conformation of ethylenediamine ring in hexametallocycle apparently is the main cause of the low site symmetry 1-C$_i$ of the central entity in the crystal structure of bis[N-(2-hydroxyethyl)-ethylenediamine]copper(II) chloride [18] consisting of [Cu(hencd)$_2$Cl$_2$] units (Fig. 4.2.10), mutually bonded by hydrogen bonds.

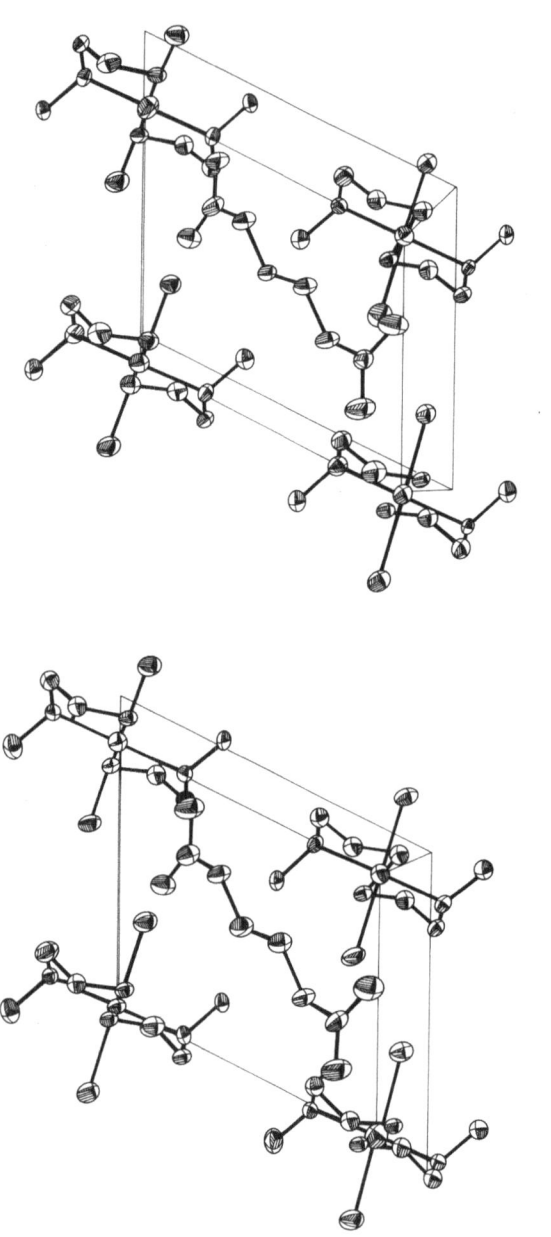

Fig. 4.2.8. Stereoview of the unit cell packing of structure [Cu(Meen)$_2$(H$_2$O)$_2$]$^{2+}$[adipate]$^{2-}$ [16]. Axis a is horizontal and axis c is vertical. Space group P$\bar{1}$, triclinic, $Z = 1$, $a = 7.241(2)$, $b = 8.271(3)$ and $c = 8.304(3)$ Å, $\alpha = 89.92(3)°$, $\beta = 114.87(2)°$ and $\gamma = 90.45(2)°$; $R = 0.035$.

These interactions between structure units apparently are responsible also for the deviations of bond angles N(1)-Cu-Cl and N(2)-Cu-Cl from their tetragonal values. These angles are 89.7(2) (2×) and 97.1(2)° (2×). The coordination of Cu^{2+} is distorted octahedral. The asymmetry of hened ligands manifests itself also in a small but significant difference of bond lengths Cu-N(1) and Cu-N(2) with the values of 2.018(5) (2×) and 2.059(2) Å (2×). The rocked conformation of ethylenediamine ring and the asymmetry of hencd moiety is probably responsible for the deviation

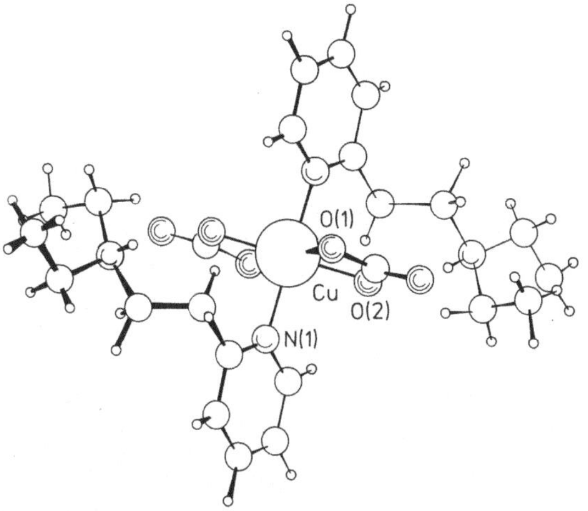

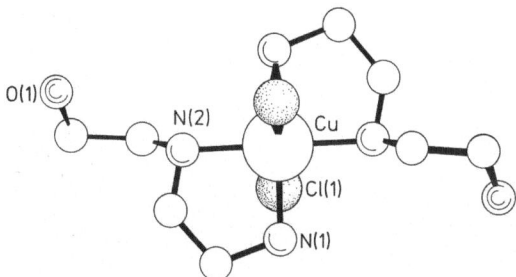

Fig. 4.2.9. Structure unit of [Cu(DEAEPH)$_2$(NO$_3$)$_2$] [17]. Space group P$\bar{1}$, triclinic, $Z = 1$, $a = 8.320(6)$, $b = 8.639(7)$, $c = 11.10(1)$ Å, $\alpha = 102.38(5)°$, $\beta = 99.95(7)°$ and $\gamma = 89.85(7)°$; $R = 0.037$.

Fig. 4.2.10. Structure unit of [Cu(hencd)$_2$Cl$_2$] [18]. Space group P$\bar{1}$, triclinic, $Z = 1$, $a = 6.569(3)$, $b = 9.472(5)$ and $c = 6.158(3)$ Å, $\alpha = 98.07(3)°$, $\beta = 101.57(3)°$, $\gamma = 80.76(4)°$; $R = 0.071$.

of bond angle N(1)-Cu-N(2) from 90° to the value of 86.3(2)° (2×). The axial bond length of Cu—Cl of 2.831(2) Å (2×) is apparently stabilized by intermolecular hydrogen bonds as also by close packing of [Cu(hencd)$_2$Cl$_2$] structure units in the crystal.

In spite of a possible higher symmetry of chromophore type CuN$_2$O$_2$Cl$_2$ site symmetry of the central entity is lowered to that of $\bar{1}$-C$_i$ owing to two asymmetrical ligands in the crystal structure of dichlorodiaquobis(dicyandiamide)copper(II) [19]. The structure units of [Cu(dcd)$_2$(H$_2$O)$_2$Cl$_2$] (Fig. 4.2.11) are mutually held by a system of hydrogen bonds shown in Fig. 4.2.12. Entity Cu^{2+} shows a distorted octahedral coordination by two nitrogen atoms from dcd ligands and two oxygen atoms from water molecules being mutually in *trans*-positions.

The lifting of degeneration of electron states of central monatomic entity by asymmetrical caffeine ligand is apparent in the crystal structure of aquadichlorocaffeinecopper(II) [20]. The structure consists of chains of mutually connected moieties

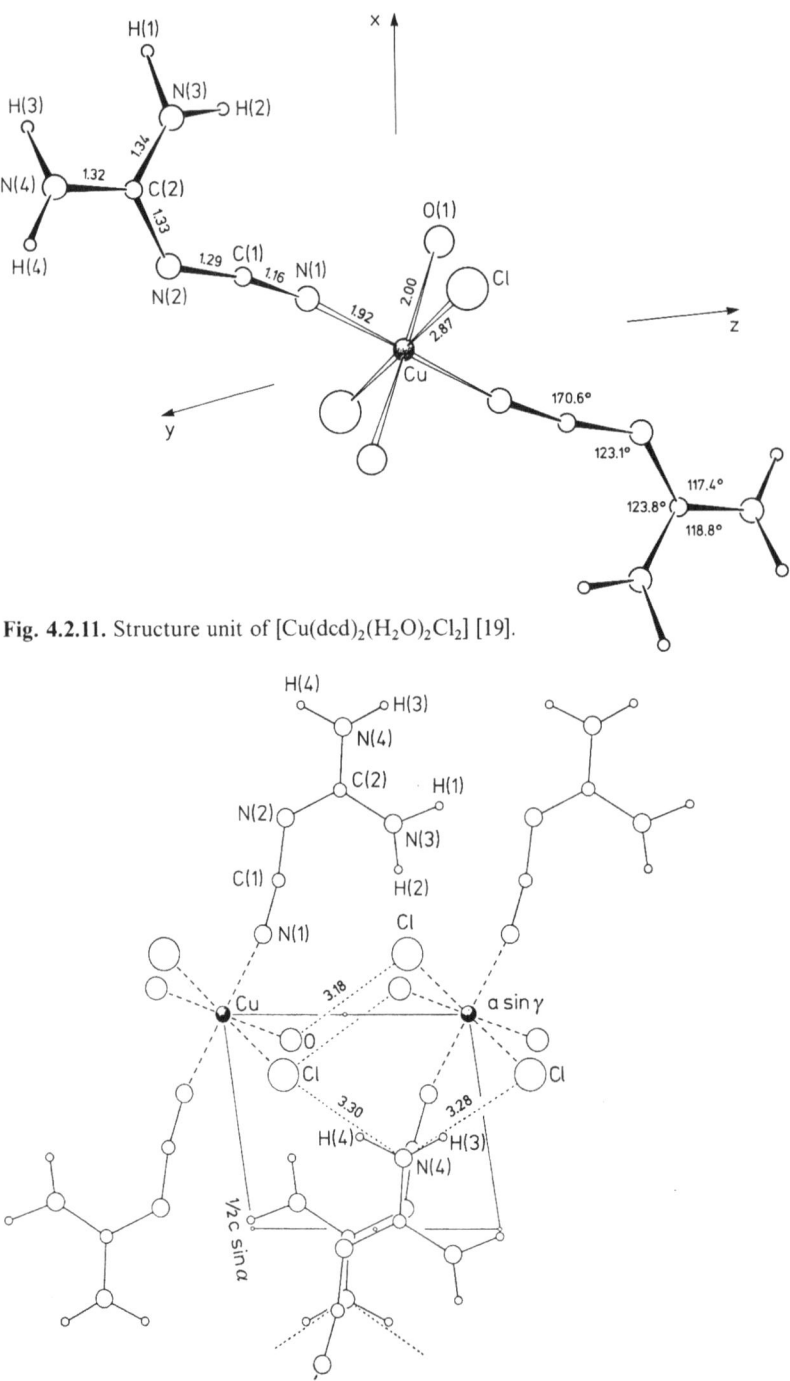

Fig. 4.2.11. Structure unit of [Cu(dcd)$_2$(H$_2$O)$_2$Cl$_2$] [19].

Fig. 4.2.12. Projection of the crystal structure of [Cu(dcd)$_2$(H$_2$O)$_2$Cl$_2$] along [010] [19]. Space group P$\bar{1}$, triclinic, $Z = 1$, $a = 5.42(1)$, $b = 6.45(1)$, $c = 9.31(1)$ Å, $\alpha = 74.5(2)°$, $\beta = 80.4(3)°$ and $\gamma = 84.7(3)°$; $R = 0.083$.

[CuCl$_2$(caf) (H$_2$O)] (Fig. 4.2.13) with comparatively weak interactions. Entity Cu^{2+} is then pentacoordinated with axial bond length Cu-Cl(1)' 2.788(2) Å. Atoms Cl(1), N(9), Cl(2), however, significantly deviate from the least squares plane. The Cu^{2+} entity is shifted out of this plane by 0.15 Å in direction to Cl(1)' atom. The axial bond Cu-Cl(1)' is normal to the basal plane (angle 89.9°). The bond lengths Cu-Cl(1), Cu-N(9), Cu-Cl(2) and Cu-O(10) are 2.319(2), 1.98(1), 1.96(1) and 2.248(2) Å, respectively. The least squares plane of imidazole ring of otherwise significantly convex caf ligand, forms, with the basal plane of the distorted tetragonal pyramid as coordination polyhedron around the central entity, an angle of 87.7°. Thus the hexacoordination of Cu^{2+} is rendered impossible and to the detriment of Cu-Cl(1) bonds infinite chains are formed along axis 2$_1$. The site symmetry of Cu^{2+} entity is lowered to the lowest possible symmetry viz. 1-C$_1$, and thus entity Cu^{2+} can occupy the site of point configuration, which is realized only in the Weissenberg lattice complex P2$_1$2$_1$2$_1$(a). Coordination of Cu^{2+} is tetragonal pyramidal, distorted towards trigonal bipyramidal. The chains in the crystal are mutually kept by hydrogen bonds of type Cl ... HOH.

By the trivariant lattice complex of P2$_1$2$_1$2$_1$(a) it can be described that relative positions of Cu^{2+} entities of the crystal structure of copper(II) chelate glycyl-L-leucyl-L-tyrosine [21]. Asymmetrical glt moiety is chelate bonded to the central entity under the formation of an infinite chain of dimers Cu(1) (glt)$_2$Cu(2), being joined together

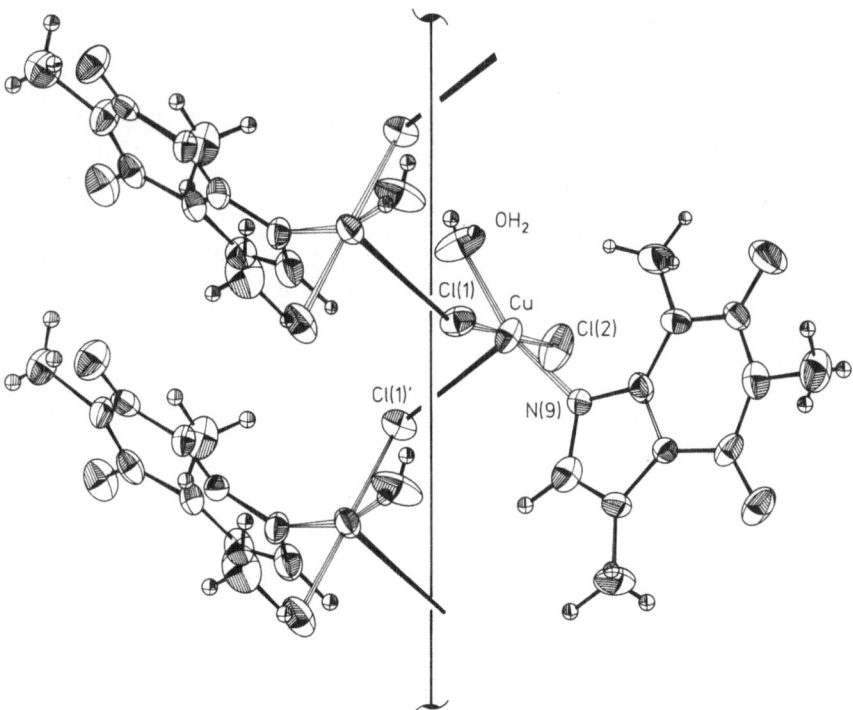

Fig. 4.2.13. View of polymeric chain in the structure of [CuCl$_2$(caf) (H$_2$O)] [20]. Space group P2$_1$2$_1$2$_1$, orthorhombic, $Z = 4$, $a = 16.370(9)$, $b = 13.432(7)$ and $c = 5.814(6)$ Å; $R = 0.028$.

via free oxygen atoms of glt moieties by O-Cu bonds linked into infinite chains along axis 2_1 (Fig. 4.2.14). In this way the central entity can occupy only the site of symmetry $1\text{-}C_1$. The approximately tetragonal-pyramidal coordination of Cu(1) and Cu(2) is formed by five membered chelate rings of glt moieties creating a base. The bond lengths in basal planes of Cu(1) and Cu(2) entities are in the ranges of 1.92(2)–2.00(2) and 1.94(2)–2.02(2) Å. The bond lengths Cu(1)-O(1) and Cu(2)-O(2) in axial directions are 2.57(2) and 2.32(2) Å. The central entities are somewhat displaced from the basal planes in direction to the apices of the tetragonal pyramids. The chains are mutually kept together by hydrogen bonds between the remaining monodentate oxygen atoms and bidentate nitrogens of peptide moieties and also by hydrogen bonds type peptide-peptide.

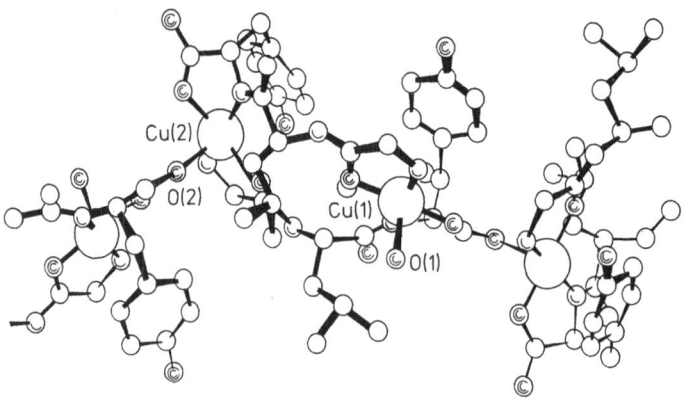

Fig. 4.2.14. Structure of chains in crystal structure of $Cu_2(glt)_2.8H_2O.C_4H_{10}O$ [21]. Space group $P2_12_12_1$, orthorhombic, $Z = 4$, $a = 9.316(2)$, $b = 25.76(2)$ and $c = 21.05(1)$ Å; $R = 0.103$.

The expressively asymmetrical surroundings shows Cu^{2+} entities also in the crystal structure of dichloro[2-(2-methylaminoethyl)pyridine]copper(II) [22], consisting of chains of $[Cu(MAEP)Cl_2]_n$ units mutually joined by bridges-Cu-Cl-Cu-along the screw axis 2_1. Atoms Cl(1), N(2), Cl(2) and N(1) forming the base of the distorted tetragonal pyramid as the coordination polyhedron around Cu^{2+} do not lie precisely in the plane. The first two of them are above the plane of least squares by 0.10 Å, and the remaining are below the plane also by 0.10 Å. The central entity is displaced from basal plane by 0.16 Å in direction of Cl(1)′ atom (Fig. 4.2.15). The bond length Cu-Cl(1)′ makes 2.785(2) Å. The sixmembered metallocycle shows *boat* geometry. The bond angle N(1)-Cu-N(2) is 90.9(1)°. Pentaccordination of Cu^{2+} in this case is apparently stabilized by the formation of chains along axis 2_1, to the detriment of bridges Cu-Cl-Cu, whereby the degeneration of electronic states of the central entity is removed.

Hexacoordinated Cu^{2+} entity with tetradentate glutamate moieties creates in the structure of copper glutamate dihydrate [23] a threedimensionally infinite skeleton. The asymmetry of ligands in this case also lowers the site symmetry of the central entity so much that from the Weissenberg lattice complexes for the sites of all

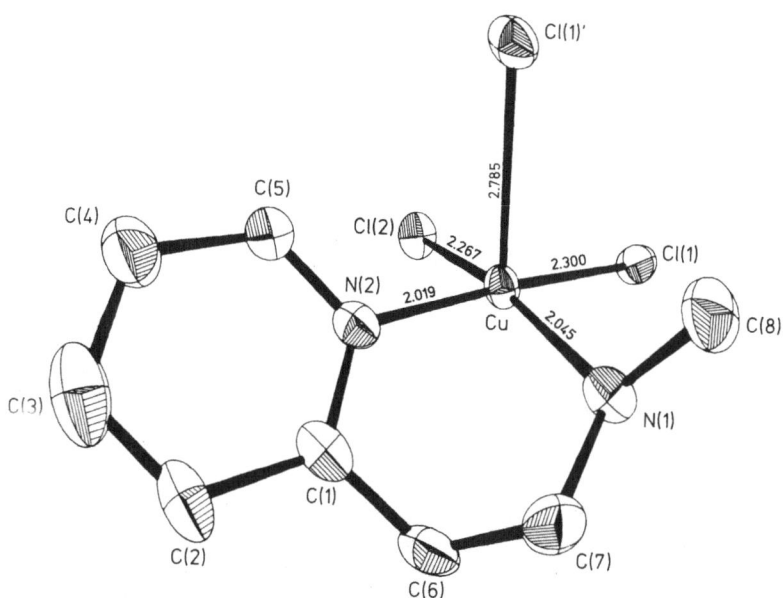

Fig. 4.2.15. Coordination geometry around Cu^{2+} in the structure of Cu(MAEP)Cl$_2$ [22]. Atom Cl(1)' is related to Cl(1) by the 2_1 screw axes parallel to c. E.S.D. of bond lengths are within 0.001–0.002 Å. Space group $P2_12_12_1$, tetragonal, $Z = 4$, $a = 19.627(9)$, $b = 8.288(3)$ and $c = 6.520(2)$ Å; $R = 0.036$.

entities that of $P2_12_12_1(a)$ is only possible. Atoms N, O(1), O(3) (Fig. 4.2.16) of two gltm moieties and atom O(5) of the water molecule form an approximately square-planar coordination of Cu^{2+}, being completed by the atoms O(2) and O(4) of gltm moieties to a strongly distorted octahedral coordination. The bond lengths in equatorial plane Cu-N(1), Cu-O(1), Cu-O(3) and Cu-O(5) are 1.996(4), 1.967(4), 1.981(4) and 1.991(4) Å, respectively. Axial bonds Cu-O(2) and Cu-O(4) have the lengths 2.299(4) and 2.588(4) Å. According to [24] the last value can be considered for bond length in the case of chromophore CuO$_6$. It can, however, be expected also in the present case of the nonhomogeneous inner coordination sphere, that this bond is also a significant stabilizing factor of the whole skeleton structure. Shortening of the bonds in the equatorial plane in the case of chromophore CuO$_6$ leads to elongation of bonds in an axial direction and vice-versa [24]. The steric possibilities of the carboxylate group of gltm ligand apparently require a lengthening of the axial bond Cu-O(4). The mean value of bond lengths in plane [25] is $R_S = 1.98$ Å and the mean value of bonds out of plane $R_L = 2.44$ Å, complying quite well to the correlation of R_L vs. R_S [24]. The system of hydrogen bonds between gltm moieties through uncoordinated water molecules is also probably a factor contributing considerably to the stability of the structure skeleton.

In the discussed crystal structures the degeneration lifting of electronic states of the central monatomic entity is always caused by the asymmetry of coordinated ligands. The hexacoordinated entity of Cu^{2+} is always surrounded by four ligand atoms with shorter bond lengths Cu-L and by two considerably longer bonds in axial directions.

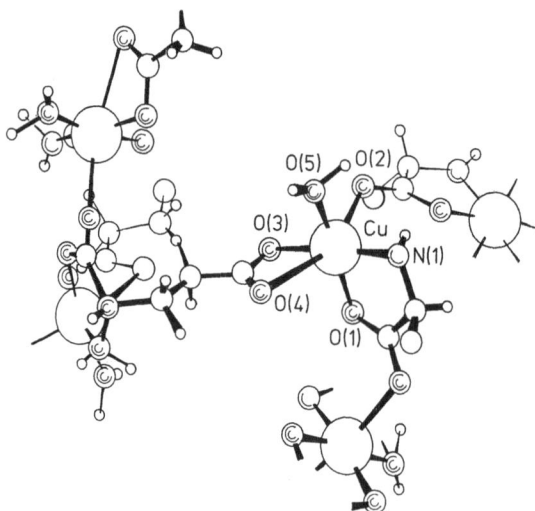

Fig. 4.2.16. Part of the skeleton in the crystal structure of Cu(gltm).2H$_2$O [23]. Space group P2$_1$2$_1$2$_1$, orthorhombic, Z = 4, a = 11.084(6), b = 10.350(3) and c = 7.238(2) Å, R = 0.032.

In cases of pentacoordinated entity, Cu^{2+} is bond Cu-L with the ligand atom in the fifth position of coordination which is significantly longer than bond lengths in the basal plane. The mean value of bond lengths in-plane is 2.02 ± 0.08 Å, this value being near to the middle of class of the greatest maximum (1.975 Å) in the Cu^{2+} histogram. The second maximum apparently corresponds to the elongated axial bonds of preferred geometries of coordination polyhedra around the central entity of Cu(II) complexes. Thus the crystallographical non-rigidity of entity Cu^{2+} is prevalingly caused by its non-rigid behavior in the inner coordination sphere. This behavior is described in [25–31] for various types of chromophores.

4.3 Ag(I) Compounds

Entity Ag^{1+} is distinguished by its great non-rigidity (Fig. 2.2.4) and by an expressive maximum in the histogram of interatomic vector lengths (p. 43) with the class middle of 2.175 Å. In agreement with its great non-rigidity this entity occupies preferentially asymmetrical sites of the Weissenberg lattice complex P2$_1$2$_1$2$_1$(a). Several crystal structures of coordination compounds of Ag^{1+} exhibit significant deviations of some bond length Ag^{1+}-O or Ag^{1+}-N from their mean value.

The crystal structure of diimidazole silver(I) nitrate [32] consists of cation and anion units [Ag(im)$_2$]$^+$ and of NO$_3^-$ mutually held by ionic and hydrogen bonds (Fig. 4.3.1). The bond lengths Ag—N in cation units are 2.120(8) and 2.132(8) Å, thus showing a negligible difference. The bond angle N-Ag-N in this unit is, however, 172.0(3)°, being a significant deviation from the linear value. Entity Ag^{1+} is deviated from the plane of one imidazole ring by 0.132 Å and by 0.217 Å from the plane of the other. A mutual angular displacement of imidazole rings around Ag-N bonds and the above-mentioned deviation of angle N-Ag-N can, from

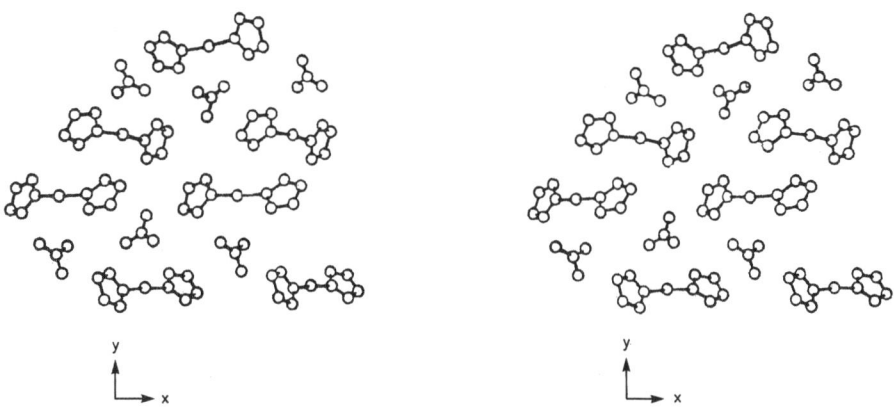

Fig. 4.3.1. Stereoscopic view of the structure Ag(im)$_2$(NO$_3$) [32]. Space group P2$_1$2$_1$2$_1$, orthorhombic, $Z = 4$, $a = 10.927(2)$, $b = 18.215(5)$ and $c = 4.999(1)$ Å; $R = 0.039$.

linearity, be explained by repulsion between cation units in **Y**-direction (Fig. 4.3.1). Such a geometry of cation units apparently is stabilized by hydrogen bonds in direction **Z** between two oxygen atoms of planar NO$_3^-$ unit and by two nitrogens from two imidazole rings of adjacent cation units.

The binuclear complex of benzoate silver(I) shows a crystal structure [33] consisting of discrete, approximately planar [Ag$_2$(benz)$_2$] units, mutually held by Van Der Waals forces. The Ag—Ag distance in the binuclear unit makes 2.902(3) Å. Each Ag^{1+} entity of this unit connects two carboxylate oxygen atoms from two benzoate ligands. Bond lengths Ag-O have a mean value of 2.225 $\pm$ 0.009 Å, while bond angles O-Ag-O are 157.0(7) and 160.3(7)° (Fig. 4.3.2). Significant differences between bond lengths O-C, C-C and between several bond angles show an asymmetry of binuclear unit, probably caused by their close packing.

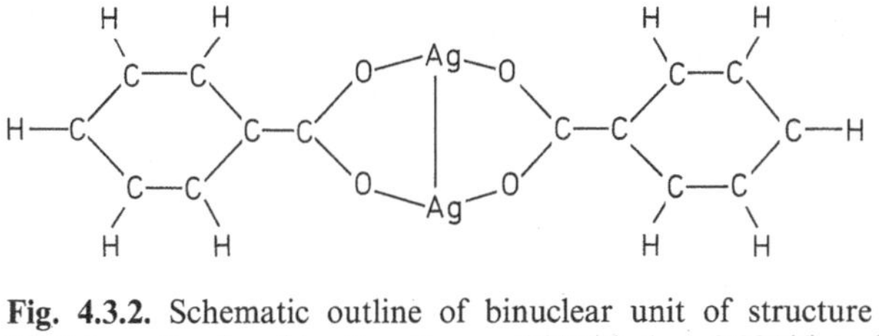

Fig. 4.3.2. Schematic outline of binuclear unit of structure Ag$_2$(benz)$_2$ [33]. Space group P2$_1$2$_1$2$_1$, orthorhombic, $Z = 2$, $a = 6.297(5)$, $b = 8.987(6)$ and $c = 23.77(2)$ Å; $R = 0.067$.

The hexanuclear complex of [Ni(δ-HCQD)$_2$Ag]$_3$.2.5CHCl$_3$ [34], where δ-HCQD$^-$ means δ-camphorquinone dioxime, consists of discrete structure units (Fig. 4.3.3) mutually held together by Van Der Waals forces. CHCl$_3$ molecules occupy interstitial sites. Entity Ag(2) is surrounded by Ni(1), Ni(2) and Ni(3) entities approximately in the apexes of the equilateral triangle. Entity Ag(2) is in plane with N(1), N(2) and N(3) and the mean value of their interatomic distances makes 3.6 Å.

Entities Ag(1) and Ag(3) are approximately equidistant with nickel entities in the distance of 4.7 Å. The distances of Ag(2)-Ag(1) and Ag(2)-Ag(3) are 3.059(5) and 3.052(7) Å. The angle of Ag(1)-Ag(2)-Ag(3) has 178.1(3)°. Thus in Ni_3Ag_3 cluster entity Ag(2) is surrounded by three nickel and two silver entities approximately in the apexes of a trigonal bipyramid. The distances between some entities, however, do not show any intermetallic bond. The ideal trigonal-bipyramidal arrangement of entities in Ni_3Ag_3 cluster apparently would correspond to the minimum of energy of electrostatic interactions of Ni-Ni, Ni-Ag and to the linear arrangement of Ag(1)-Ag(2)-Ag(3). This geometry is however, slightly distorted, influenced by δ-

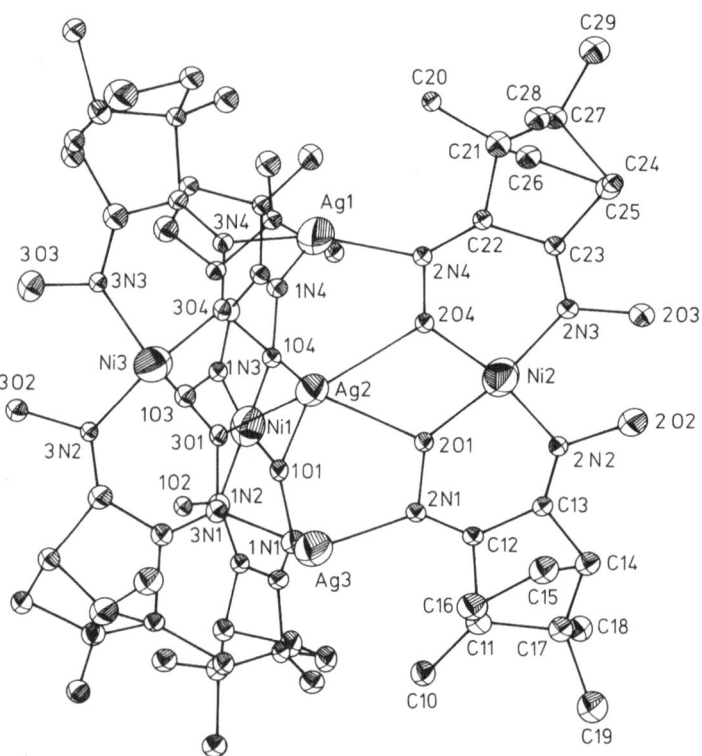

Fig. 4.3.3. The structure of unit $[Ni_3(\delta\text{-HCQD})_6Ag_3]$ [34] with 50% probability elipsoids for heavy atoms but 10% for light atoms. For clarity only the C atoms of Ni2 unit are labelled but the adopted numbering scheme is also applied to the other two units. Space group $P2_12_12_1$, orthorhombic, $Z = 4$, $a = 15.990(5)$, $b = 38.44(1)$ and $c = 13.437(5)$ Å; $R = 0.107$.

HCQD ligands forming four chelate rings around Ag(2) (Fig. 4.3.3). The structure unit of $Ni_3(\delta\text{-HCQD})_6Ag_3$ thus loses even its minimum symmetry expressed by threefold axis in the direction Ag(1)-Ag(2)-Ag(3). Dihedral angles between least squares planes Ag(2)-1O(4)-Ni(1)-1O(1), Ag(2)-2O(4)-Ni(2)-2O(1) and Ag(2)-3O(4)-Ni(3)-3O(1) thus are 130.2, 126.7 and 104.3° instead of the ideal value of 120°. The mean bond length Ag-O is 2.49 $\pm$ 0.03 Å.

The above-mentioned examples of Ag^{1+} complexes show how the linear coordination of Ag^{1+} is rocked, in the concrete examples, usually by the influence of ligands and interactions between structure units, in which the degeneration of electronic states of the central entity is lifted.

4.4 Cd(II) Compounds

The Cd^{2+} histogram (p. 43) shows the largest frequency maximum in the class with the middle of 2.275 Å which is somewhat higher than those found for the entity Ag^{1+} (2.175 Å). This fact is in agreement with the effective ionic radii of the respective entities. While the effective ionic radii of entity Cd^{2+}, for its most common coordination number six, is 0.95 Å, the effective ionic radii of entity Ag^{1+}, for its most common coordination number two, is only 0.67 Å [35]. Crystallographic non-

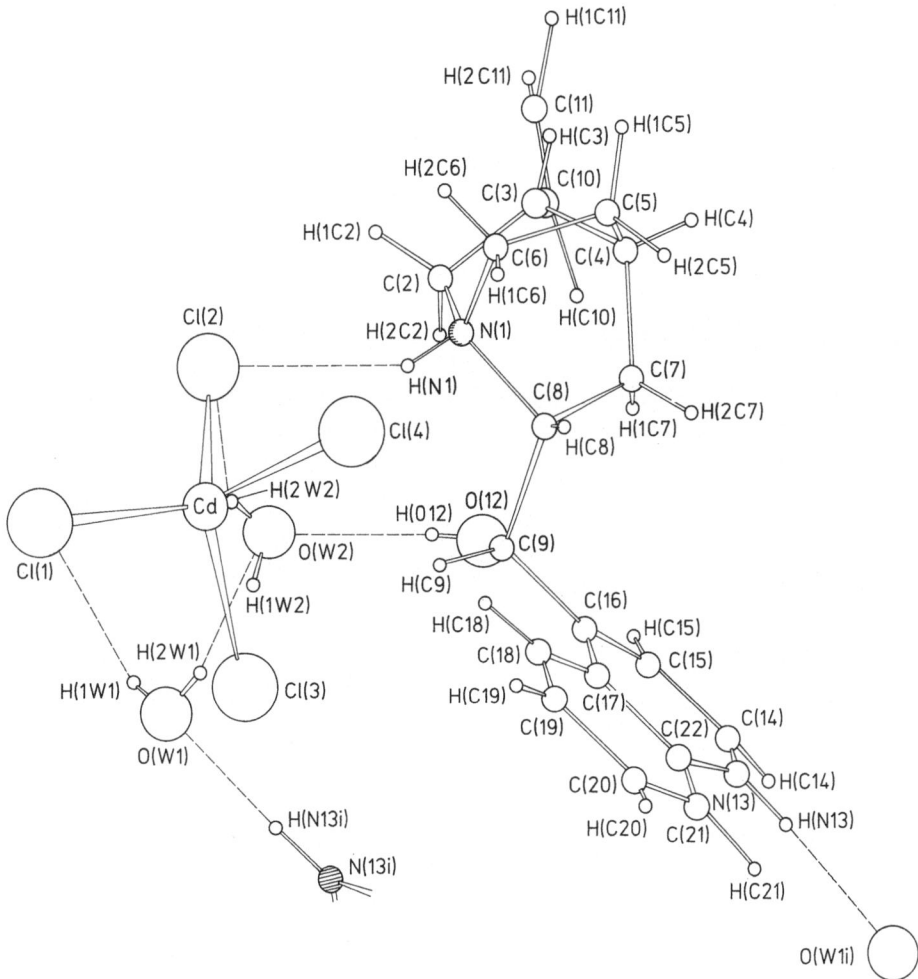

Fig. 4.4.1. The crystal structure of (cin)CdCl$_4$.2H$_2$O [36]. Space group P2$_1$2$_1$2$_1$, orthorhombic, $Z = 4$, $a = 8.1918(2)$, $b = 32.891(6)$ and $c = 8.292(2)$ Å; $R = 0.056$.

rigidity around entity Cd^{2+} is, however, bigger than those of entity Ag^{1+} (Fig. 2.2.4). Another two maxima found in the histogram of Cd^{2+} (p. 43) indicate larger variability of geometries of coordinated polyhedra about Cd^{2+} than those of Ag^{1+}.

The crystal structure of orthorhombic cinchoninium tetrachlorocadmate(II) dihydrate (cin) $CdCl_4 \cdot 2H_2O$ [36] is shown in Fig. 4.4.1. The structure consisting of the $[CdCl_4]^{2-}$ anion, the (cin)$^{2+}$ cation, and two water molecules. Central entity, Cd^{2+} is approximately tetrahedrally coordinated by four Cl atoms at distances 2.420(2), 2.493(2), 2.440(2) and 2.456(2) Å. The bond angles vary in the range 97.9–117.2°. The observed deviations from the ideal tetrahedral geometry seem to be connected with the fact that Cl(1) and Cl(2) are involved in hydrogen bonds as can be seen in Fig. 4.4.1. The units are linked spirally by hydrogen bonds between chlorine atoms, water molecules and protonated nitrogen atoms of cinchonium moiety.

In another orthorhombic cadmate(II) complex, $\{[Cd(cmp)(H_2O)].H_2O\}_n$ [37], the central entity, Cd^{2+} is in a distorted square-pyramidal environment (Fig. 4.4.2). The apex is occupied by N(3) of the cytosine ring with Cd-N(3) 2.32(1) Å and in the basal plane are three phosphate O donors. Bond distance Cd-O(3), Cd-O(4) and Cd-O(5) are 2.267(7), 2.21(1) and 2.248(7) Å. Bond length between the central entity and the water molecule Cd-O(W1) is 2.39(1) Å. The squarepyramidal array of donor atoms is severely distorted from its ideal geometry. The O_{eq}-Cd-O_{eq} angles in the square plane differ from the ideal square pyramidal angles of 86.6° and 151.9°. The observed values range from 78.9(3) to 83.3(3)° (average 81.1°) and 159.4(4)°, respectively. The N_{ax}-Cd-O_{eq} angle in the ideal square pyramid is 104.1°, compared to the observed range of 85.0(4)° to 127.8(4)° (average 103.3(4,24.5)°). (The first number in parenthesis is the e.s.d., and the second is the maximum deviation from the mean). The complex is threedimensionally polymeric, with cylindrical channels consisting of cross-linked -Cd-O-P-O-Cd- spirals. The structure is influenced by the hydrogen bonds about the uncoordinated water molecule. Elongation of Cd-Cl(3) bond length as well as deformation of the O_{eq}-Cd-O_{eq} bond angles are given by an existence of the infinite spiral chains.

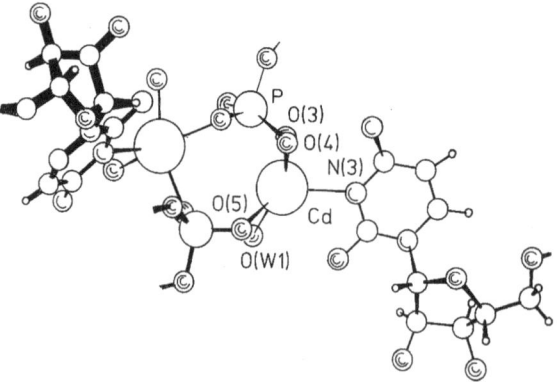

Fig. 4.4.2. Coordination of Cd^{2+} in the crystal structure $\{[Cd(cmp)(H_2O)].H_2O\}_n$ [37]. Space group $P2_12_12_1$, orthorhombic, $Z = 4$, $a = 5.294(1)$, $b = 17.070(1)$ and $c = 16.371(1)$ Å; $R = 0.038$.

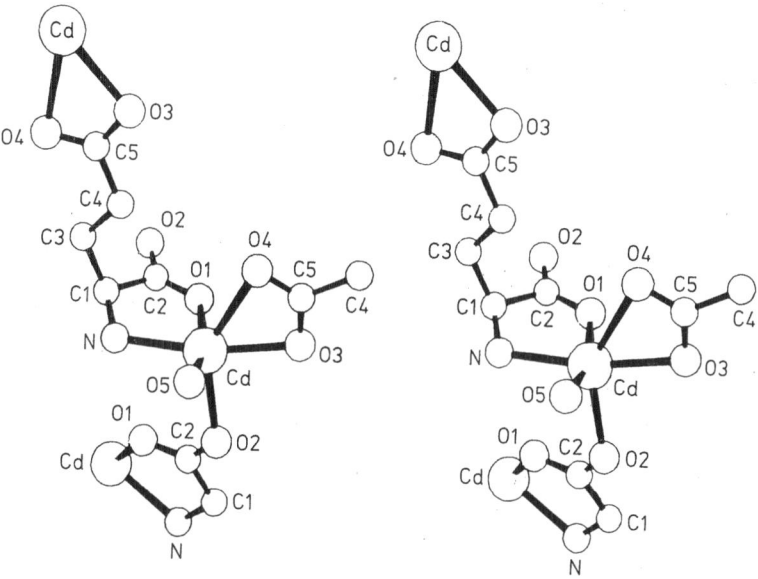

Fig. 4.4.3. Stereoscopic diagram of bonding in Cd(L-gln) (H₂O).H₂O [38].

The colourless prismatic crystals of Cd(L-glu)H₂O · H₂O show a distorted octahedral configuration around cadmate(II) [38] which is shown in Fig. 4.4.3. The central entity Cd^{2+} is coordinated by one glutamate ligand via the N(amino) (2.299(5) Å) and one O(1) (carboxyl) atom (2.288(5) Å), by a second glutamate *via* the other O(2) (carboxyl) of the terminal carboxyl group (2.252(5) Å), by a third glutamate via both O(3) and O(4) (carboxyl) atoms of the ride-chain carboxyl group (2.317(5) and 2.458(5) Å), and by a water molecule (Cd-O(5) = 2.258(5) Å). The central entity is displaced 0.22 Å from N, O(1), O(3) and O(5) plane towards O(2). Structure of the skeleton is stabilized also by a system of hydrogen bonds of uncoordinated water molecules (Fig. 4.4.4). Distortion of coordination polyhedra about Cd^{2+} is in this case apparently given by a bond, possibilities of L-glu ligand and by the arising of the infinite threedimensional skeleton.

Hydrogen bonding and packing viewed down the *a* axis of Cd(ac)₂ · 2H₂O [39] is shown in Fig. 4.4.5. As it can be seen the central entity Cd^{2+} is in a distorted square base-trigonal cap environment. Both acetate groups are bidentate, although one of the oxygen atoms, O(1), is also in a bridging position forming a continuous Cd—O spiral around the two fold screw axis parallel to *c* (Fig. 4.4.5). There are two distinct Cd-O distances, involving both acetate groups: Cd-O(2) 2.294(4), Cd-O(1) 2.597(4), and Cd-O(3) 2.304(4), Cd-O(4) 2.546(4) Å. This may be expected for the acetate group containing the bridging oxygen, O(1), (Fig. 4.4.5). The anomaly which is also to be found in long and short Cd-O bonds associated with the acetate group which is only chelating, was discussed [39]. The "bite" of the chelating acetate ligand ca. 2.2 Å is relatively rigid and can be a geometry-determining factor, as well as ligand-ligand repulsions and packing forces. Both O(3) and O(4) are involved in fairly hydrogen bonds, i.e. O(3) ... O(5) 2.75 Å, and O(4) ... O(6) 2.69 Å, and the closest non-bonded interactions of these atoms are O(3) ... O(1)

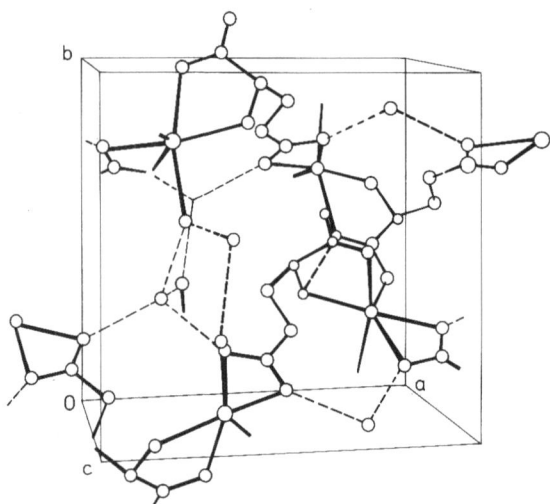

Fig. 4.4.4. Molecular packing in the unit cell of Cd(L-gln)H$_2$O.H$_2$O [38]. Space group P2$_1$2$_1$2$_1$, orthorhombic, $Z = 4$, $a = 11.61(1)$, $b = 10.79(1)$ and $c = 7.286(7)$ Å; $R = 0.038$.

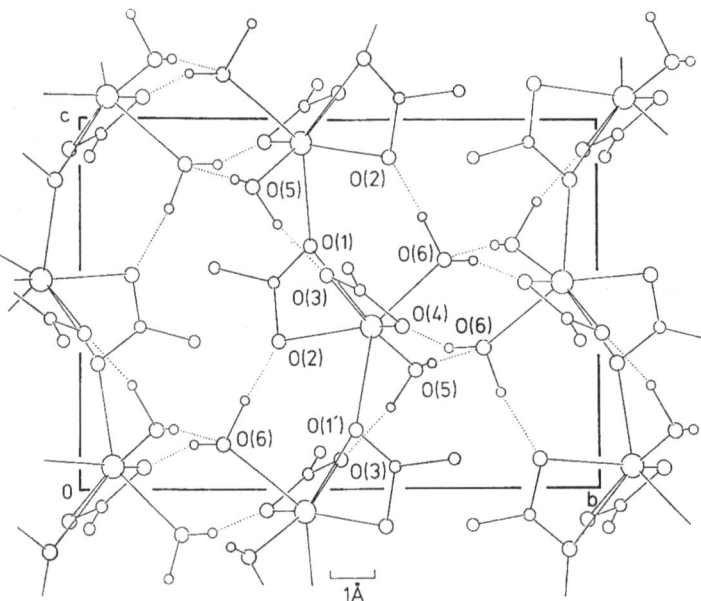

Fig. 4.4.5. Hydrogen bonding and packing viewed down the **a** axis of Cd(ac)$_2$.2H$_2$O [39]. Space group P2$_1$2$_1$2$_1$, orthorhombic, $Z = 4$, $a = 8.690(2)$, $b = 11.920(2)$ and $c = 8.100(2)$ Å; $R = 0.028$.

▶

Fig. 4.4.6. A stereoscopic view of the structure CdO(ac)$_2$.2H$_2$O [40] with the hydrogen bonds marked by thin lines. Space group P2$_1$2$_1$2$_1$, orthorhombic, $Z = 4$, $a = 7.3934(7)$, $b = 8.895(1)$ and $c = 13.354(2)$ Å; $R = 0.054$.

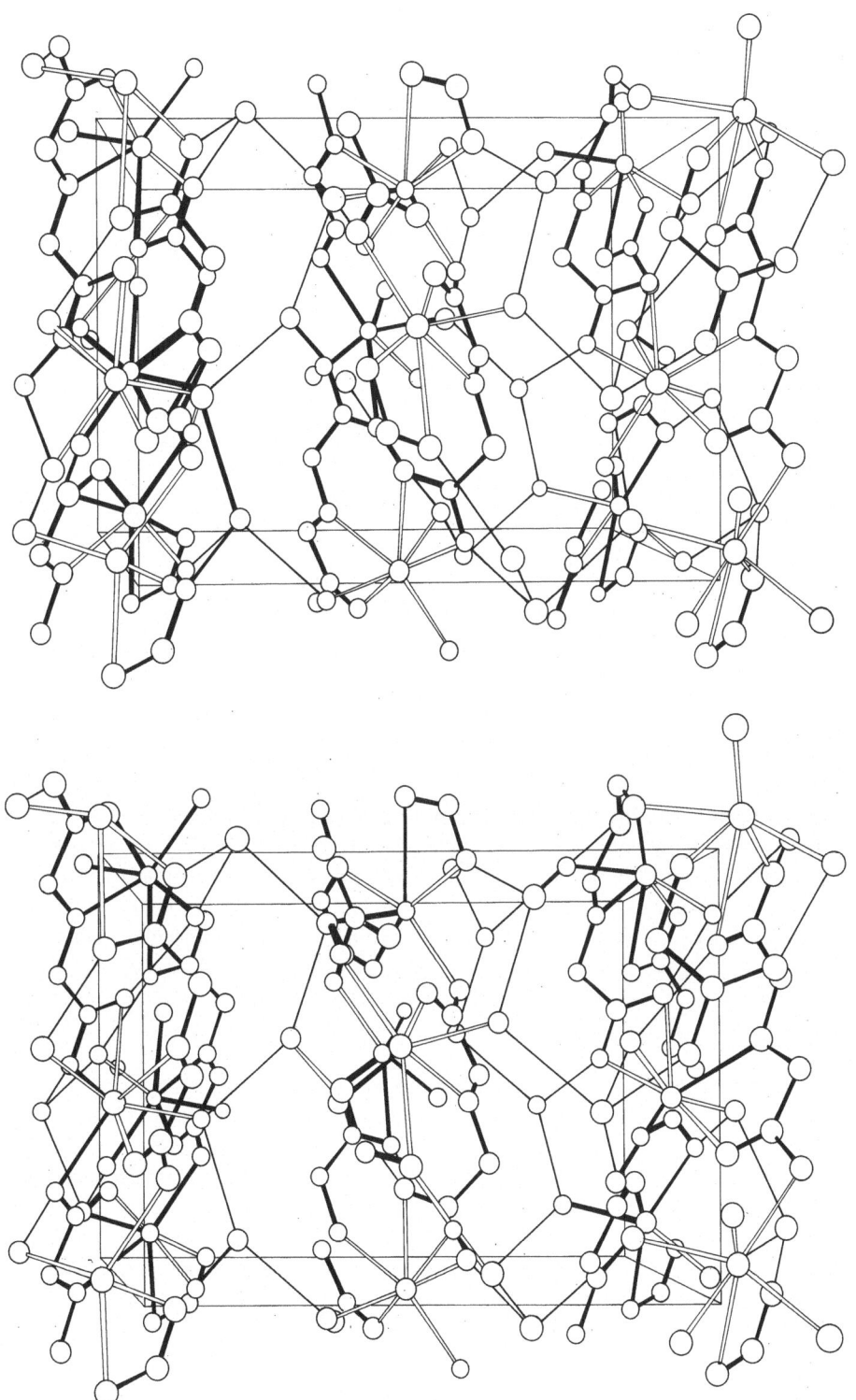

and O(3) ... O(2), 3.15 and 3.24 Å, and O(4) ... O(6) and O(4) ... O(1'), 3.01 and 3.27 Å.

Finally, in the series of cadmate(II) complexes the colourless prismatic crystals of $CdO(ac)_2 \cdot 3H_2O$ [40] consists of $CdO(ac)_2 \cdot H_2O$ layers joined by hydrogen bonds via water molecules (Fig. 4.4.6). Each central entity, Cd^{2+} is surrounded by ligands in the form of a distorted pentagonal bipyramid, the pentagon of which consists of oxydiacetic ions (Cd—O distances ranging from 2.269(15) to 2.637(15) Å with an average 2.407(15,230) Å). The apical positions are occupied by water molecules. The apical Cd—OH_2 bond distances are 2.32(1), 2.34(1) Å, respectively. The oxydiacetate ion is twisted and has the *trans-gauche* conformation. Deviations from the

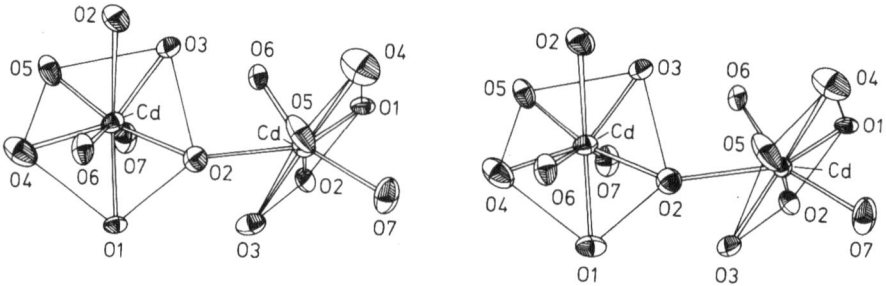

Fig. 4.4.7. The coordination around two neighbouring Cd atoms. The pentagons are marked by thin lines [40].

ideal pentagonal bipyramid geometry can be seen for example from the observed O_{ax}-Cd-O_{ax} angle of 154.2(4) Å which is much smaller than 180°. It seems that the observed deviations from the ideal pentagonal bipyramidal geometry are given mainly by mutually connected coordinated entities (Fig. 4.4.7) via an oxygen atom and by the hydrogen bond possibilities of rigid oxydiacetic anions (O(4) and O(5)). Infinite twodimensional layers are parallel to plane (001) and are held together by hydrogen bonds of coordinated and uncoordinated water molecules (O(6) and O(8)).

4.5 Fe(II) Compounds

From the coordination compounds of iron in the studied sets, compounds of Fe^{2-} and Fe^{3+} do not exhibit any significant averaging of extrinsic factors influencing interatomic vector lengths (Table 2.1.1). Fe^{2-} and Fe^{3+} histograms (p. 31 and 32) compared with those of Fe, Fe^{1+}, and Fe^{2+}, show, however, a dominant accumulation of interatomic vector lengths to higher values, outside the region of their coordination spheres. In the cases of Fe^{2-} and Fe^{3+} there is probably a more significant action of extrinsic effects. Some crystal structures of Fe^{2+} compounds with maximum site symmetry of the central monatomic entity show extreme deviations of Fe-*L* bond lengths.

Crystal and molecular structure of orthorhombic Fe(tpp) (4-Mepip) (NO) · $CHCl_3$ and triclinic Fe(tpp) (4-Mepip) (NO) has been determined [41]. In both low-spin cases, central entities, Fe^{2+}, are in distorted octahedral environments. As can be

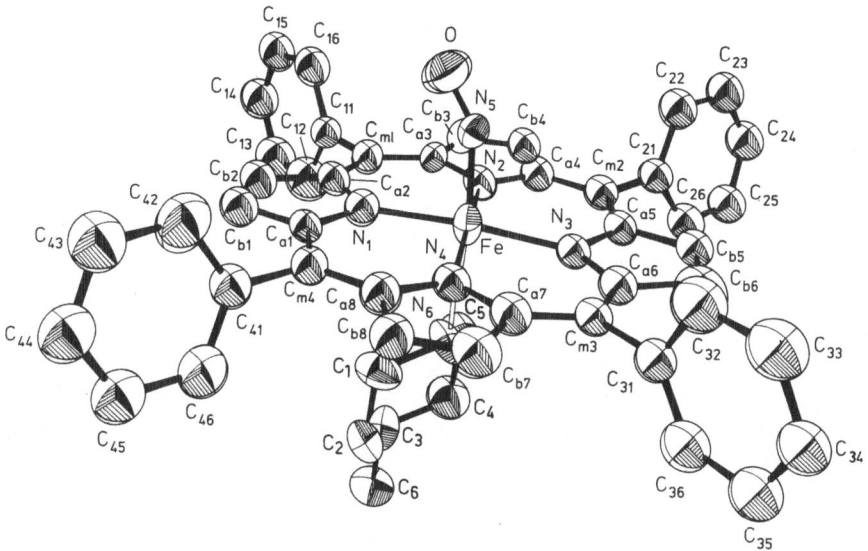

Fig. 4.5.1. Structure of Fe(tpp) (4-Mepip) (NO).$CHCl_3$ [41]. Space group $P2_12_12_1$, orthorhombic, $Z = 4$, $a = 17.638(3)$, $b = 25.582(3)$ and $c = 10.108(1)$ Å; $R = 0.109$

seen in Fig. 4.5.1, the equatorial plane is defined by the tetradentate tpp ligand, and the remaining two axial positions are occupied by NO and 4-Mepip ligands. The iron-porphinato nitrogen bond lengths are in the range of 1.991(8)–2.012(9) Å (average 2.004 (9.13) Å) in the former and 1.989(6)–2.013(6) Å (1.999(6,14) Å) in the latter. The Fe-N(NO) bond lengths are also quite similar, 1.72(1) and 1.740(7) Å, respectively. Significant differences in the two forms were noted for Fe-N(4-Mepip) bond lengths which are 2.33(1) and 2.463(7) Å, respectively. The central entities, Fe^{2+} are displaced out of plane towards NO ligand by 0.09 and 0.11 Å, respectively. The elongation of Fe-N_{ax} bond lengths is in the unsolvated form owing to larger steric interactions between axial ligand atoms and atoms of the porphinato core greater than those in the solvate form. It is presumable that the displacement of the central entity from the basal plane is caused by a certain strain in the porphinato core and by the action of packing of Fe(tpp) (4-Mepip) structure units on the methylpyridine moiety. Angular displacement of the axial Fe-N-O group around the direction of Fe-N bond as a consequence of these factors allows the displacement of the central entity out of the basal plane.

As X-ray analysis of purple high-spin Fe(tpp) $(thf)_2$ shows [42] the crystals are of triclinic space group P1, with one molecule in the unit cell ($Z = 1$) and with cell parameters: $a = 11.354(3)$, $b = 11.804(3)$ and $c = 9.688(3)$ Å, $\alpha = 103.92(2)°$, $\beta = 115.91(3)°$ and $\gamma = 102.38(2)°$. Central entity, Fe^{2+}, is in a distorted octahedral environment. As a consequence of the in-plane high-spin Fe^{2+} entity there is a larger radial expansion of the porphinato core (Fe-N 2.054(2) and 2.060(2) Å) than those in low-spin Fe(II) atom [41]. The elongation of the Fe(II)-N(tpp) bond lengths is consistent with the population of the $3d_{x^2-y^2}$ orbital in this high-spin complex [43]. The axial Fe-O(thf) bond distance of 2.351(3) Å, is consistent both with the assignment of the thf ligand as a weak field ligand and with the population of the d_{z^2} orbital as required by the high-spin ground state.

The FeN_4O_2 moiety departs significantly from the maximum possible symmetry of D_{4h}. The deviation is due to a substantial tipping of the Fe-O vectors from the perpendicular to the mean planes. The Fe-O vector is 4.1° from being perpendicular to the N_4 plane and 8.3° from the 24-atom mean plane of the core [42]. The site symmetry of the central entity thus becomes lowered. Owing to the close packing of [Fe(tpp)(thf)$_2$] structure units Fe^{2+} occupies coplanar position in tpp hole with FeN_4 plane, by which it keeps its site symmetry $\bar{1}$-C_i. Occupation of this position by entity Fe^{2+} probably also supports the centrosymmetrical arrangement of tpp ligands with respect to the geometrical centre of tpp moiety. The central entity keeps its high-spin state in spite of its coordination number six. The high radial expansion of porphynato core indicates the rise of the strain in tpp moiety.

The structure of royal-blue diaquo-2,13-dimethyl-3,6,9,12,18-pentaazabicyclo-[12,3,1]octadeca-1(18),2,12,14,16-pentaeneiron(II)chloride perchlorate was determin-

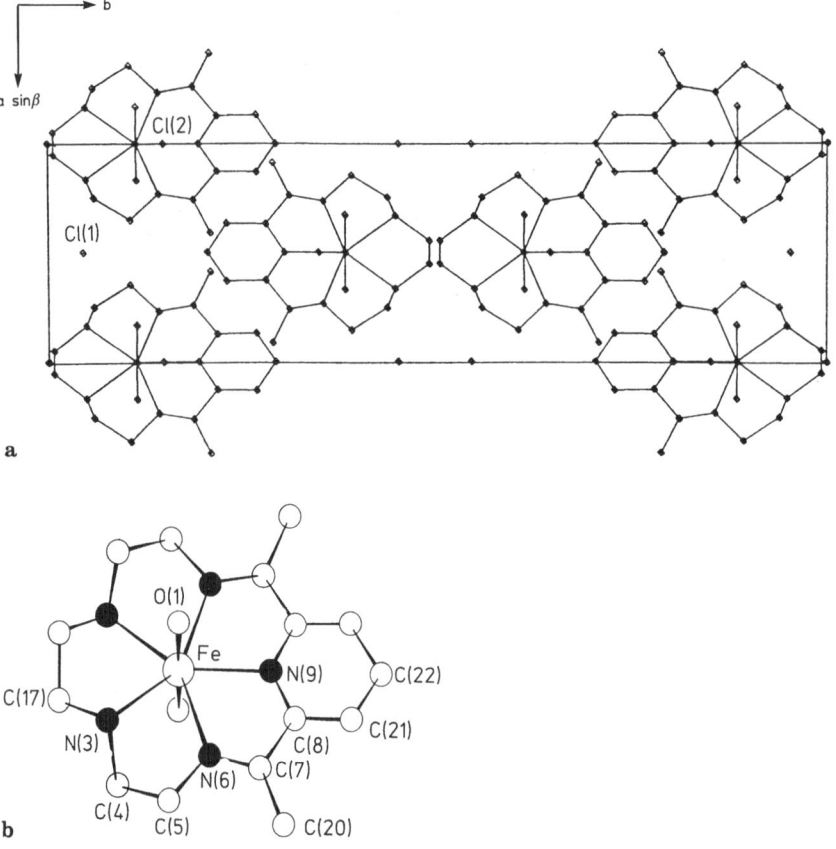

Fig. 4.5.2a, b. The unit cell in the c projection of [FeB(H$_2$O)$_2$]Cl(ClO$_4$) [44] oxygen atoms in ClO$_4^-$ ion are omitted for clarity (**a**); structure of the [FeB(H$_2$O)$_2$]$^{2+}$ cation (**b**). Space group I2/a, monoclinic, $Z = 4$, $a = 7.872(8)$, $b = 27.76(2)$, $c = 10.920(9)$ Å and $\beta = 113.6(1)°$; $R = 0.073$.

ed [44]. As can be seen on Fig. 4.5.2(b) the $FeB(H_2O)^{2+}$ cation has crystallo-graphically imposed C_2 symmetry with the central entity, Fe^{2+}, in a distorted pentagonal bipyramidal environment. The five N atoms of the macrocycle form the pentagonal girdle (Fe-N(3) 2.256(7) Å (2×), N(6) 2.262(6) Å (2×) and N(9) 2.220(7) Å) and the water molecules occupy the axial positions. (Fe-O 2.218(7) Å). The perchlorate anion is disordered. The FeN_5 atoms are not coplanar, the N atoms are displaced from the least squares plane. The unit cell is shown in the c projection on Fig. 4.5.2(a). In this case the central entity keeps its site symmetry 2-C_2 owing to symmetry of B moiety. The bond angles O-Fe-N are in the range of 89.8(1)–92(3)°. Their deviations from tetragonal values are probably caused by coulomb interactions between $[FeB(H_2O)_2]^{2+}$ units and Cl^-, ClO_4^-, and by their close packing. This influence manifests itself also in significant deviations of B moieties from planarity.

4.6 Cr(III), Mo(V), Pd(II), Co(III), Ni(II), Zn(II), Nd(III), and Hg(II) Compounds

The distortion of octahedral coordination by nitrogen atoms from ethylenediamine ligands is found in the crystal structure of high-spin complex tri(ethylenediamine)-chromium(III) Br_3.0.6hydrate [45]. The structure consists of discrete units (+)-$[Cr(en)_3]^{3+}$, of Br^- and of water molecules. The structure is also stabilized by a system of hydrogen bonds of type N-H ... O and O-H ... Br. The unit cell covers three crystallographically independent cation units (Fig. 4.6.1). The chelate-bonded ethylenediamine units with Cr(1) and Cr(2) entities are of conformations *lel lel ob* [46]. Conformations of ethylenediamine ligands chelate bonded with Cr(3) can be described as intermediates between *lel lel ob* and *lel ob ob*. The triad of complexes

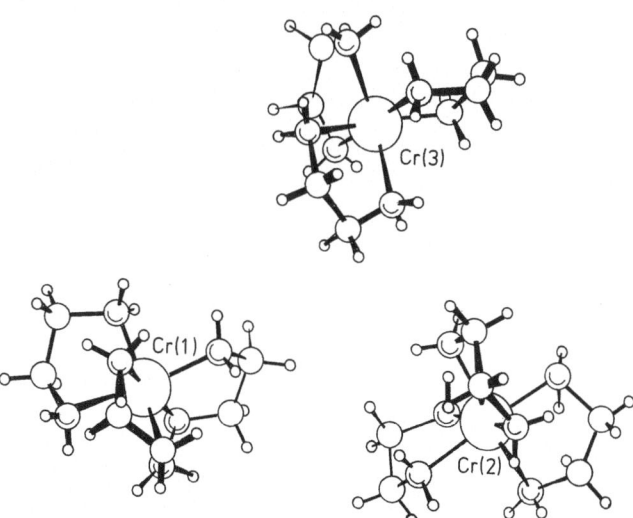

Fig. 4.6.1. Structure units in the crystal structure of (+)-$[Cr(en)_3]Br_3$.0.6H_2O [45]. Space group $P2_12_12_1$, orthorhombic, $Z = 12$, $a = 21.019(4)$, $b = 16.460(3)$ and $c = 15.346(3)$ Å; $R = 0.080$.

in Fig. 4.6.1 is of configuration $\Lambda(\delta\delta\lambda)$, $\Lambda(\delta\delta\lambda)$ and $\Lambda[\delta(\delta\lambda)\,\lambda]$. The bond lengths of Cr-N around entities Cr(1), Cr(2) and Cr(3) are in the ranges of 2.02(2)–2.15(2); 2.02(2)–2.11(2) and 2.01(2)–2.09(2) Å. Ligand conformations stabilized by electrostatic equilibrium between cation and anion units are responsible for the trigonal distortions of octahedral coordinations of Cr^{3+} entities. Such octahedron distortions can be described by means of the angular displacement φ of two opposite triangle faces of octahedron and by their interplanar distance h [47]. The angle φ expressing the measure of the octahedron distortion to the trigonal antiprism around Cr(1), Cr(2) and Cr(3) are 10(1), 7(2) and 6(2)°. Parameters h for these coordinations are 2.7(4), 4.8(4) and 1.8(4) Å.

The trinuclear high-spin complex of $[OCr_3(CH_3COO)_6.3H_2O]^+ \cdot Cl^- \cdot 6H_2O$ [48] consists of cation structure units forming a packing with channels along the twofold screw axes. These channels are filled out with disordered Cl^- anions and water molecules. Through water molecules hydrogen bonds are realized, stabilizing the pseudo-hexagonal packing of cation units. Within one trinuclear cation unit (Fig. 4.6.2) each Cr^{3+} entity is approximately octahedrally coordinated by four oxygen atoms from acetate groups, by one oxygen atom from water molecule and by one oxygen atom coordinated to all three Cr^{3+} entities of the complex. The bond lengths Cr-O around Cr(1), Cr(2) and Cr(3) are in the ranges of 1.92–2.01, 1.86–2.04 and 1.86–2.02 Å. The distortions of octahedrons around Cr^{3+} entities are apparently caused, above all, by steric hindrances of rigid acetate groups and by coulomb acting between cation and anion units. The distortions caused by the influence of the last factors manifest themselves mainly in axial bond angles $O(1)$-Cr-OH_2. These bond angles around Cr(1), Cr(2) and Cr(3) are 177.1, 177.9 and 176.6°.

The crystal structure of $Na_2[Mo_2O_4(cys)_2].5H_2O$ is shown in Fig. 4.6.3. Two tridentate cysteine ligands, (Mo-O = 2.269(6) and 2.309(6) Å; Mo-N = 2.223(5) and 2.226(5) Å; Mo-S = 2.50(1) Å (2×)), two *cis* terminal oxo ligands (Mo-O — 1.693(6) and 1.692(5) Å) and two bridging oxygen atoms (Mo-O — 1.934(6)–1.979(6) (average 1.956(6,23) Å)), are coordinated with the two central entities, Mo^{5+}, to form two distorted octahedra sharing a common edge of O atoms. The bridging oxygen atoms bring the central entities within 2.576(5) Å with Mo-O-Mo angles of 83.4(6,2)°.

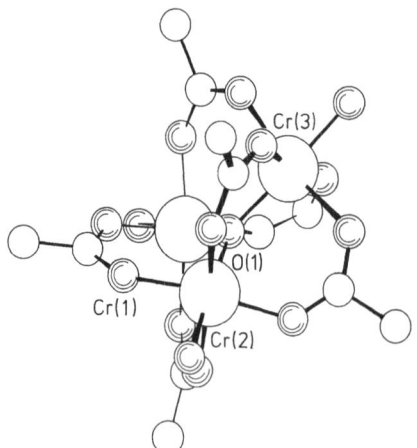

Fig. 4.6.2. Cationic structure units of $[OCr_3(CH_3COO)_6.3H_2O]^+Cl^-.6H_2O$ [48]. Hydrogen atoms are omitted. Space group $P2_12_12_1$, orthorhombic, $Z = 4$, $a = 13.68(1)$, $b = 23.14(2)$ and $c = 9.142(7)$ Å; $R = 0.128$.

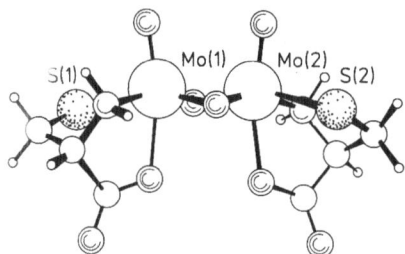

Fig. 4.6.3. The structure of $[Mo_2O_4(cys)_2]^{2-}$ [49]. Space group $P2_12_12_1$, orthorhombic, $Z = 4$, $a = 14.835(7)$, $b = 19.448(5)$ and $c = 6.491(2)\,Å$; $R = 0.052$.

X-ray analysis of red crystals of $Mo_2S_4(Et_2dte)_2$ shows [50] that each central entity, Mo^{5+} is bound to two bridging sulfur atoms, two sulfur donor atoms from the diethyldithiocarbamate ligands, and one terminal sulfur atom (Fig. 4.6.4). The Mo-Mo distance of 2.814(1) Å is somewhat longer than that of 2.576(5) Å found in $[Mo_2O_4(cys)_2]^{2-}$ anion [49]. Each molybdenium entity is displaced from the basal plane of its square pyramid towards the apical sulfur ligand by 0.73 and 0.72 Å, respectively. The dihedral angle between planes formed by S_{br}-Mo-S_{br} is 147.9(2)°. The Mo-S (terminal) bond distances, 2.090(3) and 2.094(3) Å are shorter than those of Mo-S (bridge) which are in the range 2.298(4)–2.230(4) Å (average 2.309(4,11) Å). The Mo-S(cys) bond distances range from 2.450 to 2.459(4) Å (average 2.453(4,6) Å).

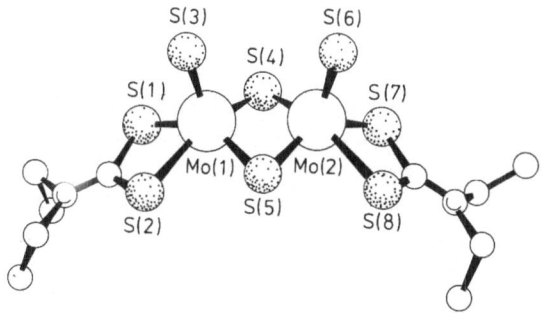

Fig. 4.6.4. Perspective view of $[Mo_2S_4(Et_2dte)_2]$ [50] with hydrogens omitted. Space group $P2_12_12_1$, orthorhombic, $Z = 4$, $a = 10.528(2)$, $b = 13.795(5)$ and $c = 14.728(2)$ Å; $R = 0.040$.

The molecular structure of $Mo_2S_4(Et_2dtc)_2$ was determined also by *Müller* et al. [51]. Their results agree with those reported by Huncke and Enemark [50].

Typical descent site symmetry of the central monatomic entity to site symmetry $\bar{1}$-C_i can be observed in the crystal structure of dichlorobis(cyclohexanone oxime)-palladium(II) [52]. The structure of this high-spin complex consists of discrete units of $Pd(cho)_2Cl_2$ mutually held by hydrogen bonds of -N-O ... Cl type. The distorted square-planar coordination is within one structure unit (Fig. 4.6.5) also stabilized by intermolecular hydrogen bonds of the said type. The bond lengths Pd-N and Pd-Cl are 2.08 and 2.24 Å. The bond angle N-Pd-Cl makes 86°.

Similarly, in the crystal structure of Pd(II) with S-methyl-L-cysteine sulfoxide [53] the asymmetrical site symmetry of Pd^{2+} entity within structure unit of $[Pd(SOMC)Cl_2]$

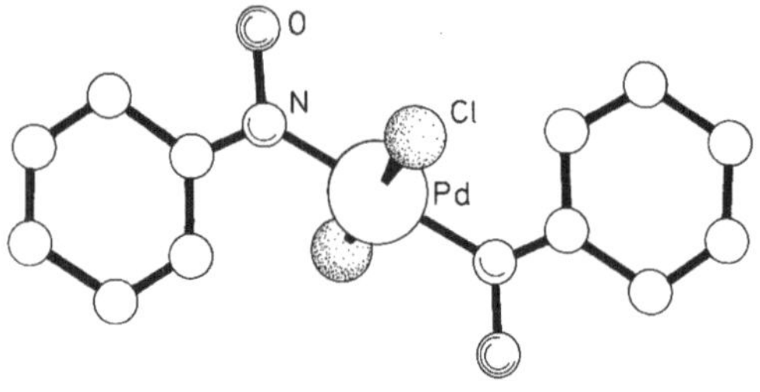

Fig. 4.6.5. The structure of unit $[Pd(cho)_2Cl_2]$ [52]. Space group P$\bar{1}$, triclinic, $Z = 1$, $a = 8.81(1)$, $b = 9.19(1)$ and $c = 4.99(1)$ Å; $\alpha = 91°32(6)'$, $\beta = 99°41(6)'$ and $\gamma = 102°02(6)'$; $R = 0.142$.

(Fig. 4.6.6) is enforced by the asymmetry of bidentate SOMC ligand. Discrete structure units are mutually connected by hydrogen bonds through the molecules of uncoordinated water. The coordination of Cd^{2+} is deviated from planarity. The bond lengths of Pd-Cl(1), Pd-Cl(2), Pd-N(1) and Pd-S(1) are 2.309, 2.296, 2.022 and 2.199 Å[1]. The distortions of the square-planar coordination around Cd^{2+} can be assigned to the rocked conformation of SOMC ligand being stabilized by hydrogen bonds of oxygen atoms of this moiety with uncoordinated water molecules

The entity of Co^{3+} in low-spin complexes is distinguished within the group of complexes d^5–d^{10} by the smallest dispersion of bond lengths Co-L [54]. Co^{3+} histogram (p. 34) also exhibits the greatest maximum in the region of the inner coordination sphere of this entity, apparently being connected with the small dispersion of Co-L lengths of this entity.

In the crystal structure of low-spin complex aquomethylbis(dimethylglyoximato)-cobalt(III) [55] the central entity, Co^{3+}, keeps its site symmetry m-C_s. The structure has orthorhombic space group Pnma, $Z = 4$ and unit cell paremeters $a = 13.18(1)$, $b = 9.115(6)$, $c = 12.132(7)$ Å and $R = 0.039$. It consists of discrete structure units of $[Co(Me_2go)_2(H_2O)(CH_3)]$ mutually bonded by hydrogen bonds. Dimethyl-

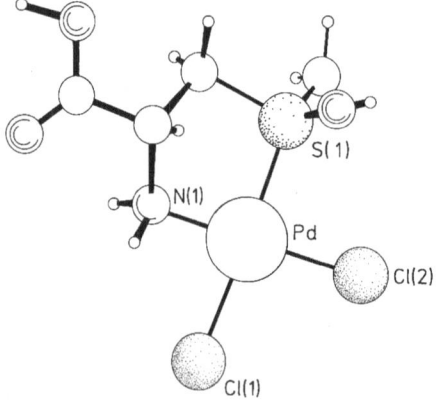

Fig. 4.6.6. Structure unit of $[Pd(SOMC)Cl_2]$ [53]. Space group $P2_12_12_1$, orthorhombic, $Z = 4$, $a = 17.858(4)$, $b = 8.690(3)$ and $c = 7.008(2)$ Å; $R = 0.061$.

---

[1] The values of bond lengths and angles were calculated by program GEOM [6].

glyoximato ligand is chelate bonded with Co^{3+} and is normally oriented to the mirror plane of the structure, forming at the same time its bisectral plane. The bond lengths Co-N are nearly equal (1.884(3) (2×) and 1.896(3) Å (2×)). Symmetrically independent bond angles N-Co-N show 98.6(1)° and 81.6(1)°. The bond lengths of Co-OH$_2$ and Co-CH$_3$ in axial direction of the slightly distorted octahedral coordination around Co^{3+} are 2.058(3) and 1.990(5) Å. Atoms O(H$_2$O) and C(CH$_3$) lie in the same mirror plane as the central entity. A higher site symmetry of Co^{3+} entity in this case is apparently made impossible by the low ligand symmetry.

Site symmetry of $2\,mm$-C_{2v} is also kept by the central entity Ni^{2+} in the crystal structure of dioxygen-bis-(t-butylisocyanide)nickel(II) [56]. Within the structure unit of [NiO$_2$(t-BuNC)$_2$] (Fig. 4.6.7), the coordination of Ni^{2+} is deviated from the square-planar form. The bond lengths of Ni-C and Ni-O are 1.840(2) (2×) and 1.808(8) Å (2×). The bond angles C-Ni-C and O-Ni-O show the values of 91.8(5)° and 47.2(4)°. The twofold axis of site group symmetry is normal to the direction of O-O bond. In the case of this high-spin complex apparently the central entity Ni^{2+} exhibits the highest possible site symmetry which still is facilitated by the coordinated ligands.

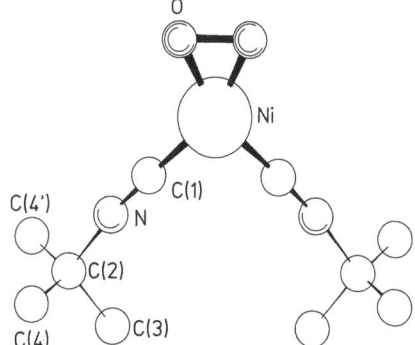

Fig. 4.6.7. Structure unit of [Ni(t-BuNC)$_2$O$_2$] [56]. Space group Cmcm, orthorhombic, $Z = 4$, $a = 11.53(1)$, $b = 16.55(2)$ and $c = 7.05(5)$ Å; $R = 0.097$.

In the crystal structure of p-aminobenzoate complex of Zn(II), [Zn(H$_2$Nbenz)$_2$]$_n$ ·1.5 nH$_2$O [57] the formation of the infinite threedimensional skeleton (Fig. 4.6.8) and steric hindrances of acetate groups from p-aminobenzoate ligands lead to site asymmetry of the central entity. The second factor apparently is the main cause of the anomalously great distance of Zn ... O(1) with the value of 2.494(8) Å. The central entity is coordinated by two nitrogen atoms (N(1) and N(2)) and two oxygen atoms (O(2) and O(3)) from different ligands. These donor atoms lie in the apexes of the distorted tetrahedron. The bond lengths Zn-N(1), Zn-N(2), Zn-O(2) and Zn-O(3) are 2.06(1), 2.053(9), 1.979(6) and 1.943(7) Å.

Quite an asymmetrical surrounding is exhibited by the central entity in the crystal structure of tri-aquo-iminodiacetato-neodymium(III) chloride [58]. Part of this structure is shown in Fig. 4.6.9. The entity of Nd^{3+} is coordinated by eight oxygen atoms and one nitrogen atom. The mean value of bond lengths Nd-O is 2.5 ± 0.1 Å and bond length Nd-N is 2.67(2) Å. Pentadentate iminodiacetate ligands link entities Nd^{3+} under the formation of infinite chains, in direction (001), whereby also the

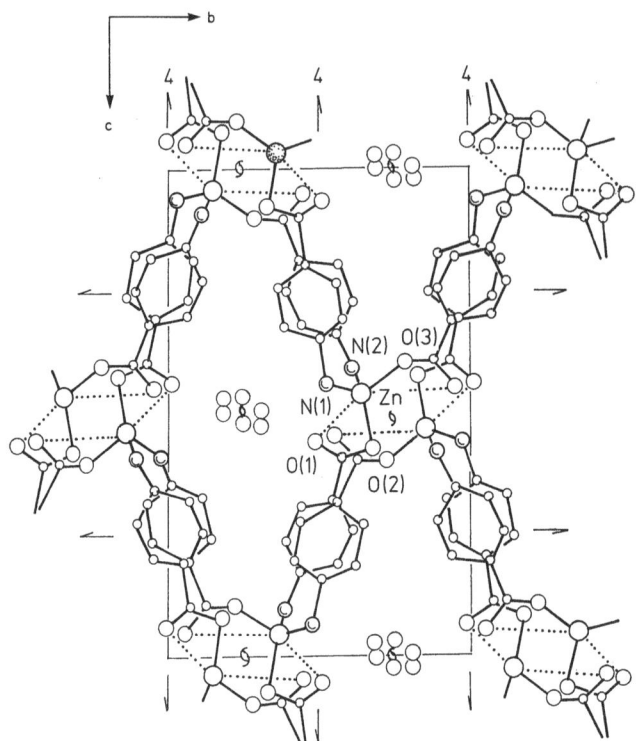

Fig. 4.6.8. Projection of the crystal structure of $[Zn(H_2Nbenz)_2]_n.1.5nH_2O$ into a plane (100) [57]. Space group $P2_12_12_1$, orthorhombic, $Z = 4$, $a = 7.623(1)$, $b = 11.189(3)$ and $c = 16.873(5)$ Å; $R = 0.052$.

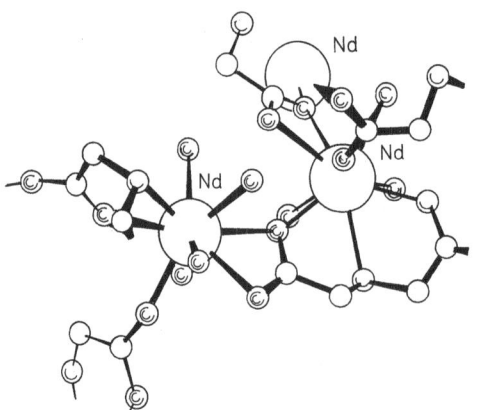

Fig. 4.6.9. Part of the crystal structure of Nd(imdac)Cl.3H$_2$O [58]. Space group $P2_12_12_1$, orthorhombic, $Z = 4$, $a = 8.356(1)$, $b = 14.164(2)$ and $c = 8.424(1)$ Å; $R = 0.087$.

asymmetrical sites of the central entities are given. The chains are mutually linked by hydrogen bonds via Cl$^-$ units.

    X-ray structure analysis of diiodo (N,N,N',N'-tetramethylthiuram monosulphide) mercury(II) [59] showed distorted tetrahedral coordination around Hg^{2+}. Within each structure unit $[I_2Hg(tmtm)]$ (Fig. 4.6.10) the central entity is coordinated by two

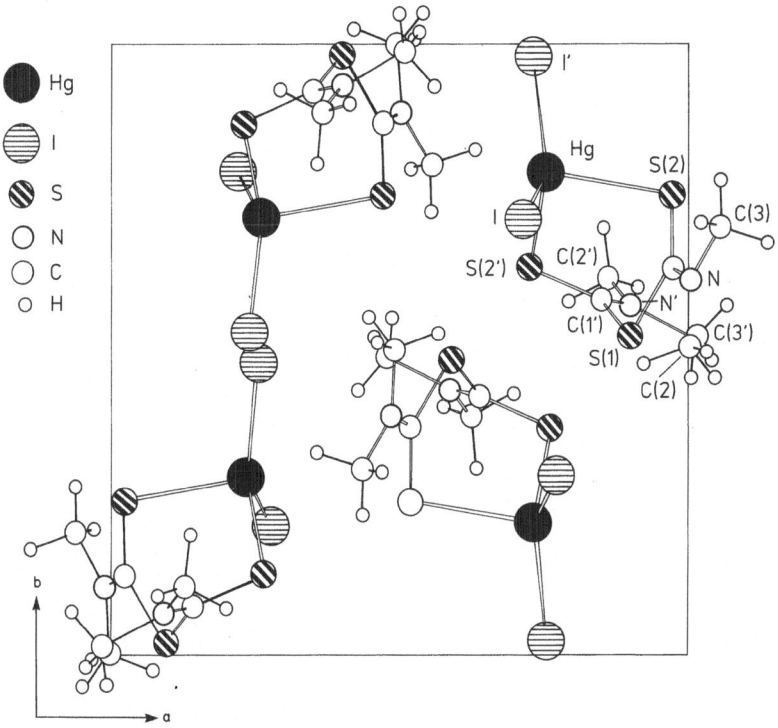

Fig. 4.6.10. Unit cell contens of $I_2Hg(tmtm)$ [59] along c, showing atom labelling. Space group $P2_12_12_1$, orthorhombic, $Z = 4$, $a = 12.910(3)$, $b = 12.441(3)$ and $c = 9.863(2)$ Å; $R = 0.038$.

sulphur atoms from the chelate bonded tmtm moiety and by two iodines. The bond lengths of Hg-I, Hg-I', Hg-S(2) and Hg-S(2') are 2.672(2), 2.656(2), 2.832(5) and 2.659(5) Å, respectively. The essential difference in Hg-S bond lengths probably is due to steric effects of tmtm moiety as it is also indicated by the central entity deviation from the plane of atoms I, I' and S(2') being nearly equidistant with it, namely 0.38 Å.

Glossary of Symbols

M-L, *R*	Bond length between the central entity and ligand atom
s	Bond valence
R	Crystallographic residual
$F_0(hkl)$	Observed structure factor
$F_c(hkl)$	Calculated structure factor
φ	Angular displacement of two opposite triangle faces of octahedron
h	Interplanar distance of two opposite triangle faces of octahedron

4.7 References

1. Pauling, L.: J. Amer. Chem. Soc. **49**, 765 (1927)
2. Pauling, L.: The Nature of the Chemical Bond, Ithaca, Cornell Univ. Press 1940[2]
3. Brown, I. D., Shannon, R. D.: Acta Crystallogr., Sect. A, **29**, 266 (1973)
4. Brown, I. D.: The bond valence method: An empirical approach to chemical structure and bonding, in: Structure and Bonding in Crystals, Vol. II, (eds.) O'Keeffe, M., Navrotsky, A., p. 1, Academic Press Inc. 1981
5. Jørgensen, Ch. K.: Rev. Chim. Min. **20**, 533 (1984)
6. Allen, F. H., Bellard, S., Brice, M. D., Cartwright, B. A., Doubleday, A., Higgs, H., Hummelink, T., Hummelink-Peters, B. G., Kennard, O., Nother Well, W. D. S., Rodgers, J. R., Watson, D. G.: Acta Crystallogr., Sect. B, **35**, 2331 (1979)
7. Bethe, H. A.: Ann. Physik **3**, 133 (1929)
8. Griffith, J. S.: J. Chem. Phys. **41**, 516 (1964) (see this reference for citations of earlier landmark papers)
9. Lever, A. B. P.: Inorganic Electronic Spectroscopy, (ed.) Lappert, M. F., p. 204, Amsterdam, London, New York, Elsevier Publ. Comp. 1968
10. Jahn, H. A., Teller, E.: Proc. Roy. Soc. **A 161**, 220 (1937)
11. Dunn, T. M., McClure, D. S., Pearson, R. G.: Some Aspects of Crystal Field Theory, p. 25, New York, Harper and Row Publishers Inc. 1965
12. Prout, C. K., Armstrong, R. A., Curruthers, J. R., Forrest, J. G., Murray-Rust, P. & Rossotti, F. J. C.: J. Chem. Soc., Sect. A, 2791 (1968)
13. Pajunen, A., Korvenrata, J.: Suomen Kemistilehti, Sect. B, **46**, 139 (1973)
14. Camerman, N., Fawcett, J. K., Kruck, T. P. A., Sarkar, B. & Camerman, A.: J. Amer. Chem. Soc., 2690 (1978)
15. Grenthe, I., Paoletti, P., Sandström, M. & Glikberg, S.: Inorg. Chem. **18**, 2687 (1979)
16. Hämäläinen, R., Turpeinen, U., Ahlgrén, M.: Acta Crystallogr., Sect B, **35**, 2408 (1979)
17. Lewis, D. L., Hodgson, D. J.: Inorg. Chem. **12**, 1682 (1973)
18. Pajunen, A., Näsäkkälä, M.: Suomen Kemistilehti, Sect. B, **45**, 47 (1972)
19. Chiesi, A., Coghi, L., Mangia, A., Nardelli, M. & Pelizzi, G.: Acta Crystallogr., Sect. B, **27**, 192 (1971)
20. Bandoli, G., Biagini, M. C., Clemente, D. A. & Rizzardi, G.: Inorg. Chim. Acta **20**, 71 (1976)
21. Franks, W. A., Van der Helm, D.: Acta Crystallogr., Sect. B, **27**, 1299 (1971)
22. Bream, R. A., Esters, E. D., Hodgson, D. J.: Inorg. Chem. **14**, 1672 (1975)
23. Gramaccioli, C. M., Marsh, R. E.: Acta Crystallogr. **21**, 594 (1966)
24. Gažo, J., Bersuker, I. B., Garaj, J., Kabešová, M., Kohout, J., Langfelderova, H., Melník, M., Serátor, M. & Valach, F.: Coord. Chem. Rev. **19**, 253 (1976)
25. Tomlinson, A. A. G., Hathaway, B. J., Billing, D. E. & Nichols, P.: J. Chem. Soc., Sect. A, 65 (1969)
26. Orgel, L. E., Dunitz, J. D.: Nature **179**, 462 (1957)
27. Gažo, J., Boča, R., Jóna, E., Kabešová, M., Macášková, Ľ., Šima, J., Pelikán, P. & Valach, F.: Coord. Chem. Rev. **43**, 87 (1982)
28. Boča, R., Pelikán, P.: Chem. Zvesti **36**, 35 (1982)
29. Boča, R.: Chem. Zvesti **35**, 769 (1981)
30. Boča, R.: Chem. Zvesti **35**, 779 (1981)
31. Hathaway, B. J., Hodgson, P. G.: J. Inorg. Nucl. Chem. **35**, 4071 (1973)
32. Antti, C. J., Lundberg, K. S.: Acta Chem. Scand. **25**, 1758 (1971)
33. Usubaljev, B. T., Movsumov, E. M., Amiraslanov, J. R., Achmedov, A. J., Musaev, A. A., Mamedov, Ch. S.: Zh. Strukt. Chim. **22**, 98 (1981)
34. Ma, M. S., Angelici, R. J., Dowell, D. & Jacobson, R. A.: Inorg. Chem. **19**, 3121 (1980)
35. Shannon, R. D.: Acta Crystallogr., Sect. A, **32**, 751 (1976)
36. Oleksyn, B. J., Stadnicka, K. M., Hodorowicz, S. A.: Acta Crystallogr., Sect. B, **34**, 811 (1978)

37. Clark, R. G., Orbell, J. D.: Acta Crystallogr., Sect. B, **34**, 1815 (1978)
38. Flook, R. J., Freeman, H. C., Scudder, M. L.: Acta Crystallogr., Sect. B, **33**, 801 (1977)
39. Harrison, W., Trotter, J.: J. Chem. Soc., Dalton Trans. 956 (1972)
40. Boman, C. E.: Acta Crystallogr., Sect. B, **33**, 834 (1977)
41. Scheidt, W. R., Brinegar, A. Ch., Ferro, E. B. & Kirner, J. F.: J. Amer. Chem. Soc. **99**, 7315 (1977)
42. Reed, A. Ch., Mashiko, T., Scheidt, W. R., Spartalian, K. & Lang, G.: J. Amer. Chem. Soc. **102**, 2302 (1980)
43. Scheidt, W. R.: Acc. Chem. Res. **10**, 339 (1977)
44. Drew, M. G. B., Bin Othman, A. H., McIlroy, P. & Nelson, S. M.: Acta Crystallogr., Sect. B, **32**, 1029 (1976)
45. Spinat, P. P., Whuler, A., Brouty, C.: Acta Crystallogr., Sect. B, **35**, 2914 (1979)
46. Harnung, S. E., Sørensen, B. S., Creaser, I., Maegaard, H., Pfenninger, U. & Schäffer, C. E.: Inorg. Chem. **15**, 2123 (1976)
47. Gagné, R. R., Koval, C. A., Smith, T. J.: J. Amer. Chem. Soc. **99**, 8367 (1977)
48. Chang, S. C., Jeffrey, G. A.: Acta Crystallogr., Sect. B, **26**, 673 (1970)
49. Liu, H., Williams, G. J. B.: Acta Crystallogr:, Sect. B, **37**, 2065 (1981); Knox, J. R., Prout, C. K.: Acta Crystallogr., Sect. B, **25**, 1857 (1969)
50. Huncke, J. T., Enemark, J. H.: Inorg. Chem. **17**, 3698 (1978)
51. Müller, A., Bhattacharyya, R. G., Mohan, N. & Pfefferkorn, B.: Z. Anorg. Allg. Chem. **454**, 118 (1979)
52. Tanimura, M., Mizushima, T., Kinoshita, Y.: Bull. Chem. Soc. Jap. **40**, 2779 (1967)
53. Allain, M., Kubiak, M., Jezowska-Trzebiatowska, B., Kozlowski, N. & Glowiak, T.: Inorg. Chim. Acta **46**, 127 (1980)
54. Valach, F., Koreň, B., Sivý, P., Melník, M.: Structure and Bonding **55**, 101 (1983)
55. McFadden, D. L., McPhail, T. A.: J. Chem. Soc., Dalton Trans. 363 (1974)
56. Matsumoto, M., Nakatsu, K.: Acta Crystallogr., Sect. B, **31**, 2711 (1975)
57. Amiraslanov, J. R., Nadzafarov, G. N., Usubaljev, B. T., Musaev, A. A., Movsumov, E. M. & Mamenov, Ch. S.: Zh. Strukt. Chim. **21**, 140 (1980)
58. Oskarsson, Å.: Acta Chem. Scand. **25**, 1206 (1971)
59. Skelton, B. W., White, A. H.: Aust. J. Chem. **30**, 1693 (1977)

5 Conclusions

Analysis of empirical distribution of interatomic vector lengths with central monatomic entities (M^{z+}) of transition elements showed their different non-rigid behavior in condensed matter. This intrinsic property of central entities can be expressed by the variance of interatomic vector lengths with common origin in the site of central entity of a certain type (M^{z+}), when the extrinsic variance (σ_g^2) has no significant contribution. It is apparent that this quantitative measure does not express completely the non-rigid behavior of central monatomic entities. In statistical approach, however, the variance of lengths of the described vectors with significantly zero extrinsic component sufficiently expresses the ability of entity M^{z+} in condensed matter to form environments with the minimum number of equidistant interatomic vectors in the sense of *VEP*. Empirical estimates of the first moments (m_1) (Table 1.1.3), of interatomic vector length distributions with significant averaging of extrinsic factors (Table 2.1.1), are in the range of 3.168 (Mn^{2+}) to 7.396 (Mn^{4+}) Å which may be considered to be the region of lengths, where the principle of vector equilibrium applies in crystal structures of coordination compounds. From the point of view of this principle (Chap. 1.2) the variance of interatomic vector lengths with insignificant extrinsic component expresses the ability of a certain central monatomic entity to deviate from the vector equilibrium.

As a suitable probability model for the distribution of interatomic vector lengths the *generalized Γ distribution* can be used, the density function of which is

$$f(x) = \frac{\lambda^{\tau}}{\Gamma(\tau)} (x - a_0)^{\tau-1} \, e^{-\lambda(x-a_0)} \qquad x \geqq a_0 \tag{5.1}$$

and $\lambda = 0.32$ Å. For monatomic entities Ti^{2+}, Mn^{1+}, Fe, Fe^{2+}, Co^{3+}, Ni^{1+}, Cu^{1+}, Nb^{5+}, Mo^{2+}, Rh, Ag^{1+}, Ce^{3+}, Nd^{3+} and Re^{3+} parameter a_0 shows the value of $a_0' = 4.25$ Å. For the other monatomic entities, the histograms of interatomic vector lengths of which presented in pp. 27–53 and $n > n_{crit}$ (Table 1.2.3), the parameters λ, τ and a_0 can be determined using a special procedure based on fitting function (5.1) with empirical distributions [1]. For vector lengths shorter than a_0' this probability model is not applicable. In this region there is probably the least contribution of the central entity to the vector equilibrium of the structure in the sense of *VEP*. The deviation of the central entity from *VEP* usually is caused by lowering its site symmetry and by simultaneously lifting the degeneration of its electron states. The statistics of occupying crystallographic sites by central monatomic entities show a lowering of their site symmetry by the *Minimax rule* [2] which means infringement of the Laves *symmetry principle* (Chap. 1.2). This is evidently also the main cause of why the central entities of transition elements only seldom occupy crystal structure sites, the symmetry of which is expressed by the main group of the corresponding space group. The selected crystal structure with such site symmetry of central monatomic entity and with extreme length of at least one bond *M-L* in the inner coordination sphere show that lowering the site symmetry of the central entity is rather effected by the influence of extrinsic factors.

According to the theorem formulated on the basis of quantum-mechanical calculations in conditions of the perturbation theory [2, 3], *ionic and covalent ligand interactions produce homomorphic Jahn-Teller distortions in electronic configurations of central entities of the type of d^n and d^{10-n}*. As the results show, between such electron-hole pairs of central entities there is no similarity in their histograms of interatomic vector lengths (pp. 27–53). Likewise, there is no legitimate significant equality between their variances (Fig. 2.2.4). In crystal structure topological distortions of the environments of central entities apparently are the results of a cooperation of extrinsic and intrinsic factors.

The influence of extrinsic and intrinsic factors on structures as a whole we will try to demonstrate with porphynate complexes of manganese. In the series of these complexes the central atom occurs in various oxidation (spin) states being stereochemically either in square-planar, or square-pyramidal or in octahedral surroundings. We shall more closely point out the individual stereochemistry and the mutual relations between the oxidation (spin) states of the central entity and the bond lengths (Mn-N) and also the mutual displacement of the central atom out of plane.

X-ray analysis of the toluene solvate of $\alpha,\beta,\gamma,\delta$-tetraphenylporphinatomanganese(II) which has been determined at 293 and 98 K [4] show that there is Mn^{2+} entity in square planar environments with out-of-plane displacement of the Mn^{2+} entity by 0.19 and 0.25 Å, respectively. The average Mn-N bond distance in this high-spin Mn^{2+} complex is 2.082 Å.

The crystal structure of Mn(tpp) (1-Meim) [5] is shown in Fig. 5.1. The manganese entity is surrounded by two ligands, in the form of a distorted square-pyramidal, with one significantly longer Mn-N (1-Meim) bond distance in the axial position (2.122(2) Å) than those in the basal plane (2.128(2.11) Å). The Mn(II) entity is displaced by 0.56 Å from the mean skeletal plane.

The mean value of the Mn(II)-N_{eq} bond distance of 2.148 (range 2.110–2.368 Å) found in high-spin manganese(II) porphyrins [5–7] is longer than those found in low-spin manganese(II) [8] and high-spin manganese(III) derivatives [9–16] which are 2.004 Å (range 2.000–2.011 Å) and 1.996 Å (range 1.952–2.022 Å), respectively. The relatively long distances in high-spin Mn(II) porphyrins are correlated with the singularly occupied $d_{x^2-y^2}$ orbital, which is unoccupied in the low-spin manganese(II) and high-spin manganese(III) derivatives.

There is a tendency of the Mn-N_{eq} bond distance to become elongated since the displacement of manganese atom from basal plane towards an axial ligand increases. The mean value of Mn-N_{eq} distances increases in the order: 1.996 Å (high-spin Mn(III) derivatives) <2.004 Å (low-spin Mn(II)) <2.148 Å (high-spin Mn(II)) and simultaneous displacements of Mn atom too, in the same order, being in the range of 0.234 to 0.356 Å high-spin Mn(III) <0.40 Å (low-spin Mn(II)) <0.56–0.730 Å (high-spin Mn(II)). Thus the geometry of the MnN_4 group is

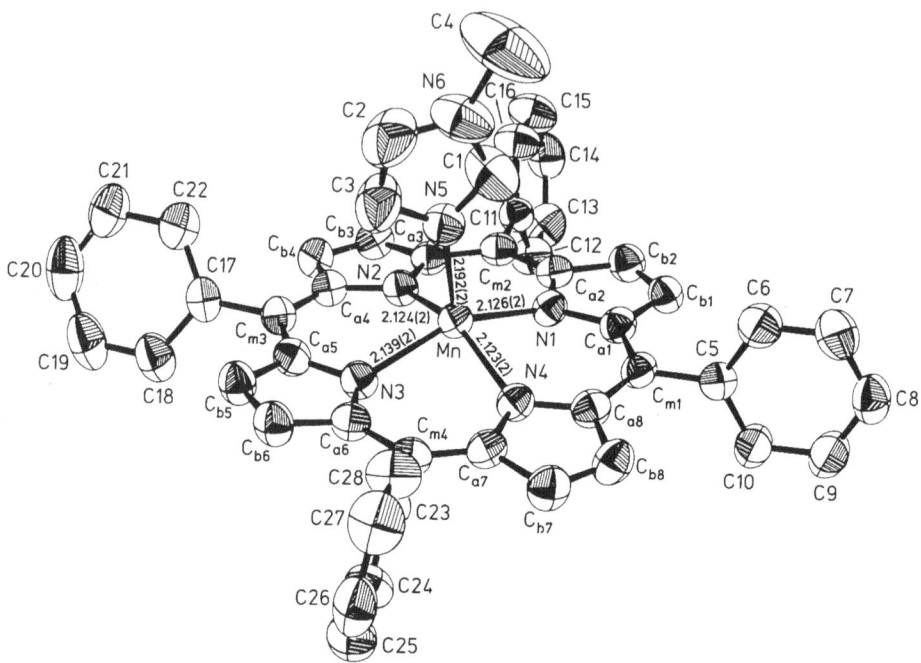

Fig. 5.1. The crystal structure of Mn(tpp) (1-Meim) [5]. Space group P2$_1$/n, monoclinic, $Z = 4$, $a = 27.405(7)$, $b = 9.645(5)$, $c = 17.768(9)$ Å and $\beta = 112.45(2)°$, $R = 0.052$.

substantially more pyramidal in the high-spin Mn(II) porphyrins than in low-spin Mn(II) and high-spin Mn(III) derivatives. The less pyramidal geometry of the MnN$_4$ group allows greater overlap between the bonding orbitals of the manganese and the tetradentate-N$_4$-ligand in high-spin Mn(III) derivatives than in low- and high-spin Mn(II) porphyrins.

In the series of hexacoordination around manganese, the plane is built up by the four nitrogen atoms of the porphyrin [8, 14–19]. The Mn-N(porphyrin) bond distances appear to show a slight decrease with an increasing formal oxidation state of manganese in the order: 2.027 Å (range 2.023–2.031 Å) for low-spin Mn(II) > 2.012 Å (range 1.991–2.046 Å) for high-spin Mn(III) > 1.991 Å (range 1.968 to 2.031 Å) for high-spin Mn(IV). The relatively short Mn-N(porphyrin) distances are correlated with an unoccupied σ-antibonding $d_{x^2-y^2}$ orbital [17].

Another fact affecting Mn-N(porphyrin) bond length is the coordination number of manganese, the former increasing, the latter increases from four to five as well as from five to six. For example, the mean Mn-N distance in four coordinated is 2.082 Å which is shorther than that of five coordinated, which is 2.148 Å; the mean Mn-N distances in five coordinated compounds 2.004 Å (low-spin Mn(II)) and 1.996 Å (high-spin Mn(III)) are shorter than those of six coordinated being 2.027 Å and 2.012 Å, respectively.

The addition of the sixth ligand leads to a substantially smaller displacement of the manganese entity from the basal plane than those of five-coordinated, which are in the range of 0.0–0.12 Å and 0.234–0.730 Å, respectvely.

From the above it is evident that the out-of-plane displacement of the central entity (manganese) in the dependence on its stereochemical arrangement increases in the order: octahedral < square-planar < square-pyramidal. This can be classified between a consequence of extrinsic factors.

Simultaneously with the increasing coordination number the bond distance Mn-N becomes elongated as one of the further manifestations of extrinsic influence.

There is a relation between the size of the ionic radius of the central atom and the bond distance Mn-N. With the increasing ionic radius of the central atom the bond distance Mn-N becomes lengthened. This may be considered an intrinsic factor. It appears, at least from the information above, that in forming the coordination polyhedron a symbiosis takes place between intrinsic and extrinsic factors, of which the determination of the dominant one is practically unambiguously not possible.

As the results of the study showed, up to the distance of 3.17 Å (m_1 value of Mn^{2+} distribution) from the central entity site, prevailingly chemical interactions may be presumed. On the other hand, however, in the range of distance of 3.17–7.40 Å, it is *VEP* that applies best. For coordination compounds in this region consisting of structure units of the type Σ_0 usually interactions of ionic, Van-der-Waals type and mediated hydrogen bonds occur. It seems that these interactions which satisfy *VEP* best are most often responsible for the existence of crystals of coordination compounds.

5.1 References

1. Valach, F., Ondráček, J., Sivý, P., Koreń, M. & Melník, M.: To be published
2. Liehr, A. D.: J. Phys. Chem. **67**, 389 (1963)
3. Liehr, A. D.: J. Phys. Chem. **67**, 471 (1963)
4. Kirnenr, J. F., Reed, Ch. A., Scheidt, W. R.: J. Am. Chem. Soc. **99**, 1093 (1977)
5. Gonzales, B., Kouba, J., Yee, S., Reed, Ch. A., Kirner, J. F. & Scheidt, W. R.: J. Am. Chem. Soc. **97**, 3247 (1975); Kirner, J. F., Reed, Ch. A., Scheidt, W. R.: J. Am. Chem. Soc. **99**, 2557 (1977)
6. Weiss, M. C., Bursten, B., Peng, S. M. & Goedten, V. L.: J. Am. Chem. Soc. **98**, 8021 (1976)
7. Anderson, O. P., Lavalle, D. K.: Inorg. Chem. **16**, 1634 (1977)
8. Scheidt, W. R., Hatano, K., Rupprecht Picinho, P. L.: Inorg. Chem. **18**, 292 (1979)
9. Weiss, M. C., Goedten, V. L.: Inorg. Chem. **18**, 274 (1979)
10. Gay, V. W., Stults, B. R., Tasset, E. L., Marianelli, R. S. & Boucher, L. J.: Inorg. Nucl. Chem. Lett. **11**, 505 (1975)
11. Tulinský, A., Chen, B. M. L.: J. Am. Chem. Soc. **99**, 3647 (1977)
12. Spreer, L. O., Malujaken, A. C., Holbrook, S. H., Otvos, J. W. & Calvin, M.: J. Am. Chem. Soc. **108**, 1949 (1986)
13. Scheidt, W. R., Ja Lee, Y., Luangdilok, W., Hallen, K. J., Anzai, K. & Hatano, K.: Inorg. Chem. **22**, 1516 (1983)
14. Day, V. W., Stults, B. R., Tasset, E. L., Day, R. O. & Marianelli, R. S.: J. Am. Chem. Soc. **96**, 2650 (1974)
15. Camenzind, M. J., Hollander, F. J. & Hill, C. L.: Inorg. Chem. **22**, 3776 (1983)
16. Kirner, J. F., Scheidt, W. R.: Inorg. Chem. **14**, 2081 (1975)
17. Camenzind, M. J., Hollander, F. J., Hill, C. L.: Inorg. Chem. **21**, 4301 (1982)
18. Hill, C. L., Williamson, M. M.: Inorg. Chem. **24**, 2836 (1985)
19. Hill, C. L., Williamson, M. M.: Inorg. Chem. **24**, 3024 (1985)
20. Scheidt, W. R., in: The Porphyrins (ed.) Dolphin, D., Vol. III, Part A, Chapter 10, New York, Academic Press, Chapter 10 and refs. cited therein

Chemical Abbreviations

ac	acetate
B	2,13-dimethyl-3,6,9,12,18-pentaazabicyclo 12,3,1 octadeca-1(18),2,12,14,16-pentaene
benz	benzoate
caf	caffein
cho	cyclohexanone oxime
cin	cinchoninium
cmp	cytidine-5′-monophosphate
cp	cyclopentadienyl
cys	cysteinate
dcd	dicyandiamide
DEAEPH	2-(2-diethylammonioethyl)pyridine
diox	dioxane
dpy	2,2′-dipyridyl
dsep	diethylselenophosphate
dtc	diethylthiocarbamate
dtp	diethylthiophosphate
en	ethylenediamine
Et_2dtc	N,N′-diethyldithiocarbamate
Eten	N-ethylethylenediamine
glt	glycyl-L-leucyl-L-tyrosine
gltm	glutamate
δ-HCQO	δ-camphorquinone dioximate
hencd	N-(2-hydroxyethyl)-ethylenediamine
his	L-histidine-D-histidine
hmpr	2-hydroxy-2-methylpropinate
H_2Nbenz	p-aminobenzoate
im	imidazole
imdac	iminodiacetate
L-glm	L-glutamate
MAEP	2-(2-methylaminoethyl)pyridine
mal	malonate
Me_2ah	3,5-dimethylaceheptylene
Meen	N-methylethylenediamine
Me_2en	N,N′-dimethylethylenediamine
Me_2go	dimethylglyoxymate

1-Meim	1-methylimidazole
2-Meim	2-methylimidazole
Mepip	4-methylpiperidine
pc	phtalocyanate
phen	o-phenanthroline
py	pyridine
SOMC	S-methyl-L-cysteinesulfoxide
t-BUNC	t-butylisocyamide
thf	tetrahydrofuran
tmtm	N,N,N′,N′-tetramethylthiuran
tpp	tetraphenylporphyrin

Appendix A

Uniform distribution (Fig. A1)

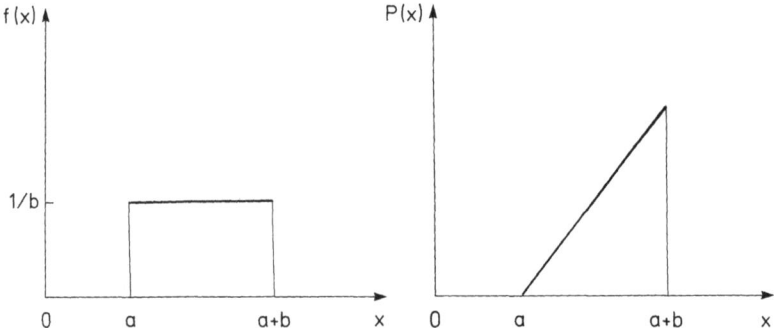

Fig. A1. Density and distribution functions of uniform distribution.

Density function: $f(x) = 1/b \qquad a \leqq x \leqq a + b$
Distribution function: $P(x) = (x - a)/b$

Γ distribution (Fig. A2)

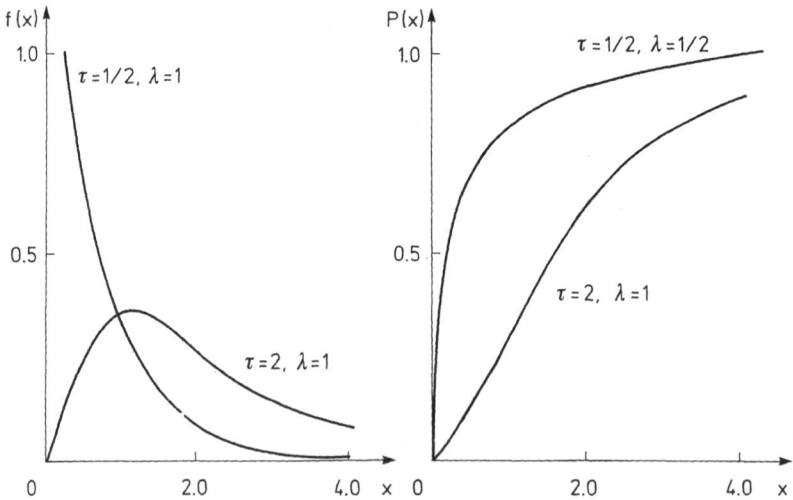

Fig. A2. Density and distribution functions of Γ distribution.

Density function:

$$f(x) = \frac{\lambda^\tau}{\Gamma(\tau)} x^{\tau-1} e^{-\lambda x} \qquad x \geq 0 \qquad \lambda > 0 \qquad \tau > 0$$

$$f(x) = 0 \qquad x < 0$$

where $\Gamma(\tau)$ is gamma-function

$$\Gamma(\tau) = \int_0^\infty x^{\tau-1} e^{-x} \, dx$$

If τ is an integer then $\Gamma(\tau) = (\tau - 1)!$
Distribution function:

$$P(x) = \frac{\lambda^\tau}{\Gamma(\tau)} \int_0^x t^{\tau-1} e^{-\lambda t} \, dt \qquad x \geq 0$$

$$P(x) = 0 \qquad x < 0$$

This function is tabulated in Reference 1.

χ^2 **distribution** (Fig. A3)

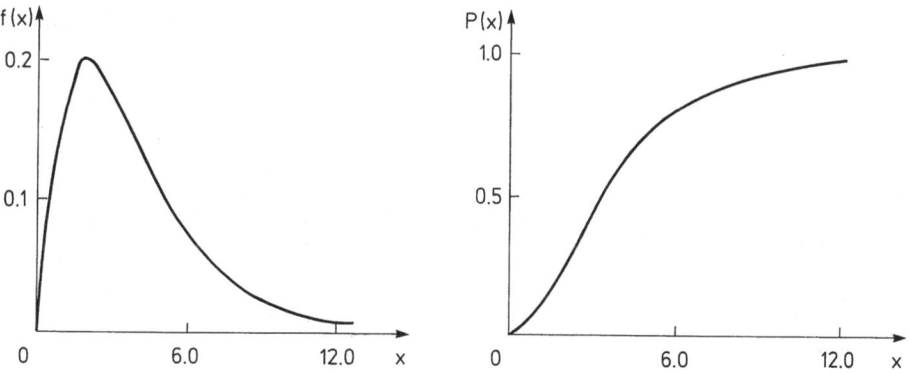

Fig. A3. Density and distribution functions of χ^2 distribution.

A special case of Γ distribution, for $\lambda = 1/2$ and τ is an integer multiple of 1/2. Parameter $v = 2\tau$ is the *number of degrees of freedom*.
Density function:

$$f(x) = \frac{x^{(v-2)/2} e^{-x/2}}{2^{v/2} \Gamma(v/2)}$$

Distribution function:

$$P(x) = \frac{1}{2^{v/2}\Gamma(v/2)} \int_0^x t^{(t-2)/2} e^{-t/2} dt$$

Exponential distribution (Fig. A4)

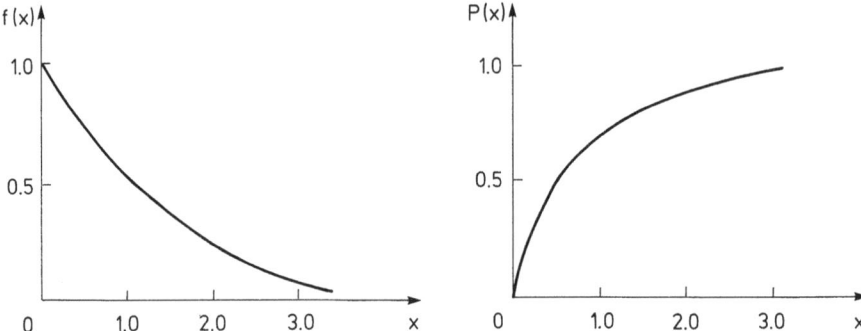

Fig. A4. Density and distribution functions of exponential distribution.

It is also a special case of Γ distribution, when $\tau = 1$.
Density function:

$$f(x) = \lambda e^{-\lambda x} \qquad x \geqq 0 \qquad \lambda > 0$$

$$f(x) = 0 \qquad\qquad x < 0$$

Distribution function:

$$P(x) = 1 - e^{-\lambda x}$$

β distribution (Fig. A5)

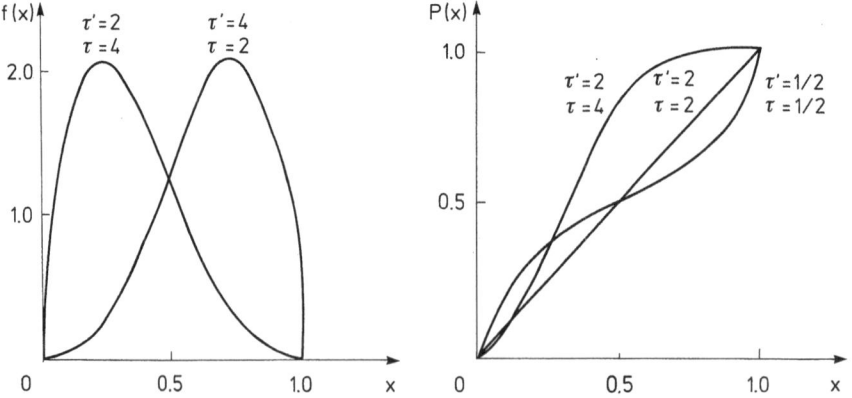

Fig. A5. Density and distribution functions of β distribution.

Density function:

$$f(x) = \frac{\Gamma(\tau + \tau')}{\Gamma(\tau)\,\Gamma(\tau')}\, x^{\tau'-1}(1-x)^{\tau-1}$$

$$0 \leqq x < 1 \qquad \tau > 0 \qquad \tau' > 0$$

Using the relation between Γ and β functions

$$\frac{\Gamma(\tau)\,\Gamma(\tau')}{\Gamma(\tau + \tau')} = \beta(\tau,\,\tau')$$

where

$$\beta(\tau,\,\tau') = \int\limits_0^1 t^{\tau'-1}(1-t)^{\tau-1}\, dt$$

we obtain

$$f(x) = \frac{x^{\tau'-1}(1-x)^{\tau-1}}{\beta(\tau',\,\tau)}$$

Distribution function:

$$P(x) = \frac{1}{\beta(\tau',\,\tau)} \int\limits_0^x t^{\tau'-1}(1-t)^{\tau-1}\, dt \qquad 0 \leqq x \leqq 1$$

Also this function is brought in [1].

Weibull distribution (Fig. A6)

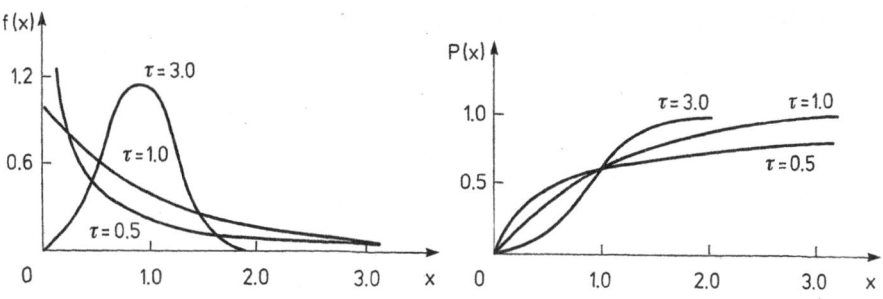

Fig. A6. Density and distribution functions of Weibull distribution.

Density function:

$$f(x) = \tau(x^{\tau-1})\,e^{-x^{\tau}} \qquad x \geqq 0 \qquad \tau > 0$$

Distribution function:

$$P(x) = 1 - \exp(e^{-x^{\tau}})$$

F distribution (Fisher-Snedecor distribution) (Fig. A7)

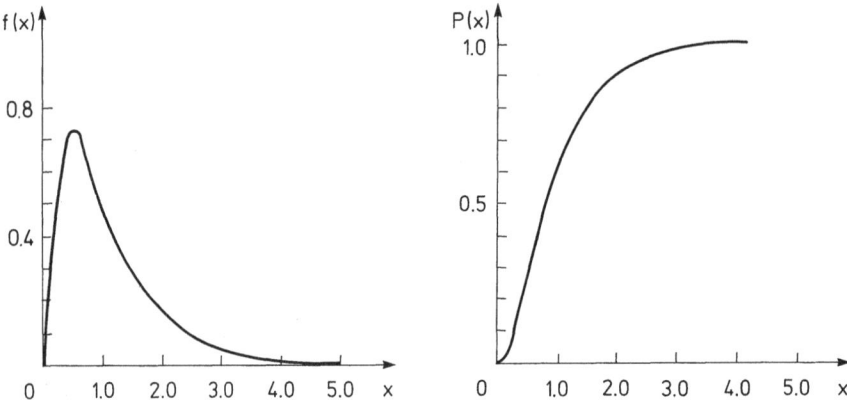

Fig. A7. Density and distribution functions of F distribution ($v_1 = 4$, $v_2 = 40$).

Density function:

$$f(x) = \frac{\Gamma[(v_1 + v_2)/2](v_1/v_2)^{v_1/2} \, x^{(v_1-2)/2}}{\Gamma(v_1/2)\,\Gamma(v_2/2)\,[1 + v_1 x/v_2]^{(v_1+v_2)/2}}$$

v_1 and v_2 are *degrees of freedom* of F distribution
The distribution function is tabulated e.g. in [2]

t distribution (Student distribution) (Fig. A8)

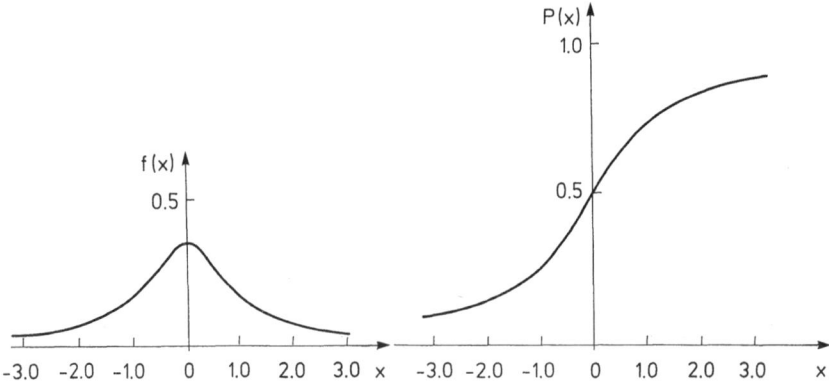

Fig. A8. Density and distribution functions of t distribution ($v = 1$).

Density function:

$$f(x) = \frac{\Gamma[(v+1)/2][1 + (x^2/v)]^{-(v+1)/2}}{(\pi v)^{1/2} \Gamma(v/2)}$$

v is the *degree of freedom* of t — distribution
This distribution function is tabulated in [2].
   The relations between the above presented distributions are described in [3].

References

1. Pearson, E. S., Hartley, H. O.: Tables for Statisticians, Vol. 1, Cambridge University Press, Cambridge 1954
2. Hald, A.: Statistical Tables and Formulas, John Wiley & Sons. Inc.
3. Hastings, N. A. J., Peacock, J. B.: Statistical Distributions, Butterworths Ltd., London 1980

Appendix B

Sample variance can be expressed

$$S^2 = \frac{1}{n-1} \sum_{i=1}^{n} (X_i - \bar{X})^2 = \frac{1}{n-1} \sum_{i=1}^{n} \left[(X_i - \mu) - \frac{1}{n} \sum_{i=1}^{n} (X_i - \mu) \right]^2$$

$$= \frac{1}{n} \sum_{i=1}^{n} (X_i - \mu)^2 - \frac{1}{n(n-1)} \sum_{i \neq j} (X_i - \mu)(X_j - \mu)$$

The expected value of this variable is

$$E(S^2) = \frac{1}{n} E\left[\sum_{i=1}^{n} (X_i - \mu)^2 \right] - \frac{1}{n(n-1)} E\left[\sum_{i \neq j} (X_i - \mu)(X_j - \mu) \right]$$

$$= \frac{1}{n} \sum_{i=1}^{n} E(X_i - \mu)^2 - \frac{1}{n(n-1)} \sum_{i \neq j} E(X_j - \mu) E(X_j - \mu)$$

$$= \frac{1}{n} n\sigma^2 - \frac{1}{n(n-1)} \cdot 0 = \sigma^2$$

Appendix C

Fourier development of function $V(r)$ may be expressed in the form of

$$V(r) = \sum_{l_1} \sum_{l_2} \sum_{l_3} \varphi(l_1 l_2 l_3) \, e^{2\pi i (l_1 x + l_2 y + l_3 z)} \tag{C1}$$

where x, y, z are fractional coordinates in the crystallographic coordinate system defined by vectors a, b, c. This coordinate system generally need not be rectangular. l_1, l_2, l_3 are integers. Cartesian coordinates generally can be converted to coordinates x, y, z by means of relations:

$$x = \alpha_{11} x_c + \alpha_{12} y_c + \alpha_{13} z_c$$

$$y = \alpha_{21} x_c + \alpha_{22} y_c + \alpha_{23} z_c$$

$$z = \alpha_{31} x_c + \alpha_{32} y_c + \alpha_{33} z_c$$

The coefficients α_{ij} depend on the angles between the axes of crystallographic and Cartesian coordinate systems. Thus there exist such vectors $v(g_1 g_2 g_3)$, by means of which the development (C1) may be written in the form of:

$$V(r) = \sum_{g_1} \sum_{g_2} \sum_{g_3} \varphi(g_1 g_2 g_3) \, e^{2\pi i v(g_1 g_2 g_3) \cdot r} \tag{C2}$$

Function $V(r)$ must fulfil the condition of periodicity with respect to vector t:

$$t = t_1 a + t_2 b + t_3 c$$

where t_1, t_2, t_3 are integers, i.e.

$$V(r + t) = \sum_{g_1} \sum_{g_2} \sum_{g_3} \varphi(g_1 g_2 g_3) \, e^{2\pi i [v \cdot (r + t)]}$$

$$= \sum_{g_1} \sum_{g_2} \sum_{g_3} \varphi(g_1 g_2 g_3) \, e^{2\pi i (v \cdot r)} \, e^{2\pi i (v \cdot t)} = V(r)$$

This condition is only fulfilled, if

$$e^{2\pi i (v \cdot t)} = 1$$

From this relation the condition follows

$$v \cdot t = v(t_1 a + t_2 b + t_3 c) = (v \cdot a) t_1 + (v \cdot b) t_2 + (v \cdot c) t_3 = \text{integer}$$

Thus scalar products in brackets also must be integers:

$$v \cdot a = g_1 \qquad v \cdot b = g_2 \qquad v \cdot c = g_3 \tag{C3}$$

Each vector v may be expressed as linear combination of three nonplanar vectors:

$$v = A(a \times b) + B(b \times c) + C(c \times a) \tag{C4}$$

Inserting this equation into (C3) leads to

$$\begin{aligned} B(b \times c) \cdot a &= g_1 \\ C(c \times a) \cdot b &= g_2 \\ A(a \times b) \cdot c &= g_3 \end{aligned} \tag{C5}$$

Let us mark the unit cell volume as:

$$\Omega_0 = (b \times c) \cdot a = (c \times a) \cdot b = (a \times b) \cdot c$$

then we will obtain from Eq. (C5) the identities:

$$B = g_1/\Omega_0 \qquad C = g_2/\Omega_0 \qquad A = g_3/\Omega_0$$

Inserting them into Eq. (C4) will give us the final form of the introduced vector:

$$v(g_1 g_2 g_3) = \frac{g_3(a \times b)}{\Omega_0} + \frac{g_1(b \times c)}{\Omega_0} + \frac{g_2(c \times a)}{\Omega_0} =$$

$$= g_1 a^* + g_2 b^* + g_3 c^* = g^*(g_1 g_2 g_3)$$

Inserting this into Eq. (C2) we shall obtain the original development in the form of:

$$V(r) = \sum_{g_1} \sum_{g_2} \sum_{g_3} \varphi(g_1 g_2 g_3)\, e^{2\pi i g^*(g_1 g_2 g_3) \cdot r}$$

Appendix D

The set $\mathbb{G}$ of elements $g_1, g_2, g_3, \ldots$ forms a *group* if the following conditions are fulfilled:

(a) To each two elements g_1, g_2 of set $\mathbb{G}$ ($g_1 \in \mathbb{G}, g_2 \in \mathbb{G}$) it exists solely one third g_3 being their product:

$$g_1 g_2 = g_3 \tag{D1}$$

and at the same time is $g_3 \in \mathbb{G}$. Thus for element g_3 it holds:

$$\Gamma'(g_3) = \Gamma'(g_1 g_2) = \Gamma'(g_1)\, \Gamma'(g_2)$$

Thus property (D1) also holds for the set of matrix representations Γ', which is the homomorphic image of $\mathbb{G}$. Representation of $\Gamma'(g_3)$ is an element of the same set as representations of $\Gamma'(g_1)$ and $\Gamma'(g_2)$

(b) The product of the group elements in associative:

$$(g_1 g_2)\, g_3 = g_1 (g_2 g_3) \tag{D2}$$

Then it holds for matrix representations:

$$\Gamma'[(g_1 g_2)\, g_3] = \Gamma'(g_1 g_2)\, \Gamma'(g_3) =$$
$$= \Gamma'(g_1)\, [\Gamma'(g_2)\, \Gamma'(g_3)] = \Gamma'(g_1)\, \Gamma'(g_2 g_3)$$

Thus the set of matrix representations also fulfils the condition of associativity (D2)

(c) There exists solely one unit element e of set $\mathbb{G}$, for which it applies

$$ge = eg = g \tag{D3}$$

Matrix representation of element g then is

$$\Gamma'(g) = \Gamma'(eg) = \Gamma'(e)\, \Gamma'(g) = \Gamma'(ge) = \Gamma'(g)\, \Gamma'(e)$$

Thus condition (D3) also applies for matrix representation in the form of

$$\Gamma'(g)\, \Gamma'(e) = \Gamma'(e)\, \Gamma'(g) = \Gamma'(g)$$

where $\Gamma'(e)$ is a *unit matrix*.

(d) To each element $\mathbf{g}$ of set $\mathbb{G}$ there exists its inverse element $\mathbf{g}^{-1}$, for which it holds

$$\mathbf{g}\mathbf{g}^{-1} = \mathbf{g}^{-1}\mathbf{g} = e \tag{D4}$$

Similarly as in the foregoing steps one may write

$$\Gamma'(e) = \Gamma'(\mathbf{g}^{-1}\mathbf{g}) = \Gamma'(\mathbf{g}^{-1})\,\Gamma'(\mathbf{g}) = \Gamma'(\mathbf{g}\mathbf{g}^{-1}) = \Gamma'(\mathbf{g})\,\Gamma'(\mathbf{g}^{-1})$$

by which the validity of condition (D4) for the set of matrix representations also is evidenced.

Subject Index